Statistical Models and Control Charts for High-Quality Processes

Statistical Models and Control Charts for High-Quality Processes

By

M Xie
Dept of Industrial and Systems Engineering
National University of Singapore
Kent Ridge Crescent, Singapore 119 260

TN Goh
Dept of Industrial and Systems Engineering
National University of Singapore
Kent Ridge Crescent, Singapore 119 260

V Kuralmani
Institute of High Performance Computing
1 Science Park Drive II, Singapore 117 528

KLUWER ACADEMIC PUBLISHERS
BOSTON/DORDRECHT/LONDON

Distributors for North, Central and South America:
Kluwer Academic Publishers
101 Philip Drive
Assinippi Park
Norwell, Massachusetts 02061 USA
Telephone (781) 871-6600
Fax (781) 681-9045
E-Mail: kluwer@wkap.com

Distributors for all other countries:
Kluwer Academic Publishers Group
Post Office Box 322
3300 AH Dordrecht, THE NETHERLANDS
Telephone 31 786 576 000
Fax 31 786 576 474
E-Mail: services@wkap.nl

Electronic Services < http://www.wkap.nl>

Library of Congress Cataloging-in-Publication Data

A C.I.P. Catalogue record for this book is available
from the Library of Congress.

Statistical Models and Control Charts for High-Quality Processes
ISBN: 1-4020-7074-8

Printed on acid-free paper.

Printed in the United States of America.

Contents

Contributing Authors

M. Xie received his PhD in Quality Technology with from Linkoping University, Sweden, in 1987. **T.N. Goh** received his PhD from the University of Wisconsin – Madison, USA, in 1973. **V. Kuralmani** received his PhD in statistics from Bharathiar University, India, in 1993. They have extensive experience in research, teaching and consulting in the area of quality engineering. Prof Xie, a Senior Member of American Society for Quality and IEEE, also serves as an editor of *Int J of Reliability, Quality, and Safety Engineering*. Prof Goh is an Elected Academician of the International Academy for Quality and a Fellow of the American Society for Quality.

Preface

Can we improve a zero-defects (ZD) process? The answer to this seemingly impossible question is actually YES! It all depends on what we mean by "zero-defect". Continuous improvement is always possible and should be both said and implemented. One should strive to "do it better each time" rather than to "do it right the first time" (that could be wrong next time). In reality, what is right today could be wrong tomorrow and vice versa. Think about Galileo and how right he was in his time, when he was regarded as totally wrong!

Statistical process control based on statistical control charts has been widely used in industry to monitor processes that are far from ZD. Its use in a near ZD manufacturing environment has faced many problems. It is our intention in writing this book to present some recently developed techniques for using control charts. The aim is to provide control charts to monitor a process in the broadest sense. This means not only process deterioration can be detected, but also possibilities of process improvement can be discovered, so that further improvement in a near ZD environment becomes possible.

It is our hope that based on rigorous analysis and the simplicity of the approaches, control charts can be better implemented especially in a high quality environment. Control charts, if properly used, can be much more useful than what has been traditionally presented. In this book, we summarize some recent findings and applications of statistical control chart techniques for processes with only occasional problems. The focus is on the modification of traditional control charts for attributes and one-at-a-time measurements. Such a process is of growing importance in industry today. Some related techniques for variable charts and charts based on conventional sampled data are also discussed.

Chapter 1 briefly reviews the control charts that are widely used today and discusses some potential problems if the traditional approach is used in a near ZD environment. We also summarize techniques for solving such problems, with details presented in the rest of the book. Notations and terminology relevant to this book are also provided although we have tried to follow the commonest ones in the existing literature.

Chapter 2 discusses the need for using probability limits, a topic that has only been occasionally discussed elsewhere. In a near ZD environment, the normal approximation is far from valid and thus probability limits can be considered to allow appropriate application of control chart principles. Chapter 3 presents a powerful technique based on counting of the cumulative conforming (*CCC*) items between two nonconforming ones. The *CCC* chart is very easy to implement, yet efficient in detecting process changes in a near ZD environment. Although *CCC* chart is developed for discrete production process in form of count of items, for continuous production, a related chart based on the cumulative quantity produced is also described. Throughout the book we pay as much attention to improvement detection as to process deterioration, and we specifically devote Chapter 4 to the issues related to process improvement. Some practical and managerial issues are discussed, as they are important to effective implementation of statistical process control.

Some extensions of the *CCC* chart with regard to its implementation are discussed in Chapter 5. The focus is the use adjusted control limits, transformed data, and a conditional procedure. The problems with inspection error and estimated control limits that might affect the effectiveness of a control chart are also discussed. Some other extensions of the *CCC* scheme are presented in Chapter 6. One is to use the cumulative number of items inspected before a fixed number of nonconforming items are observed. In this case, negative binomial distribution is used. When the process can be modeled as a ZD process subject to random disturbances which may lead to a number of defects, a modified Poisson model is developed and illustrated. Furthermore, for high-quality processes and automatically collected one-at-a-time data, some simple models for dependence are presented.

Chapters 7 to 10, which can be read independently of each other, discuss some further issues related to control chart analysis and applications. Chapter 7 presents the application of cumulative sum and exponentially weighted moving average techniques applied to *CCC*-related data. These procedures are known to be able to detect small process shifts and can be adopted with the standard *CCC* chart. In Chapter 8, some multivariate methods and their applications are studied. Economic design of control chart procedures is discussed based on the *CCC* chart in Chapter 9. Finally, Chapter 10 presents the analysis of a type of trended, but regularly adjusted process, which is common because of the widely used replacement and maintenance policies specified by equipment manufacturers. We explain how the traditional Shewhart control charts can be used with proper interpretation and how they can be adopted for other related planning and analysis.

With many examples, charts and procedures shown throughout the book, it is hoped that these seemingly advanced approaches can be adopted by practitioners. Although the readers should have a basic knowledge of statistical process control charts to appreciate the contents, little prior knowledge is needed to make use of the results.

While we have tried to be theoretically rigorous, some of the analytical proofs have been left out. However, as many references as possible are provided for those interested in exploring the details. We have also posed a number of questions and issues for further investigations. Researchers and students may find many ideas useful in their academic work. In fact, many theoretical and practical issues are worth exploring further.

M. Xie
T.N. Goh
V. Kuralmani

Singapore

Acknowledgements

Although there are three authors, the work in this book is the combined effort of a number of people. We would first of all like to thank our PhD students in the past several years (in alphabetic order): DQ Cai, B He, SL Ho, XS Lu, XY Tang and W Xie, who realized many of our ideas with their hard work. As a group, we also inspired each other with challenging questions and discussions.

We would also like to thank many researchers worldwide who have been interested in our work on high-quality process control. We should mention PLY Chan, E von Collani, E Del Castillo, FF Gan, EA Glushkovsky, K Govindaraju, F Jensen, CD Lai, DC Montgomery, H Ohta, P Ranjan, KL Tsui, W Woodall, Z Wu, ZL Yang, among others. We are also very grateful to many anonymous reviewers of our papers for their insightful comments.

We acknowledge the support from the National University of Singapore and the staff and other students in the Department of Industrial and Systems Engineering. We also appreciate the personal interest and effort by Gary Folven of Kluwer who flew all the way from Boston to Singapore to meet with us.

Finally, we would like to thank our families for their understanding and support all these years.

CHAPTER 1

Introduction

Control charts are widely used in industry as a tool to monitor process characteristics. Deviations from process targets can be detected based on evidence of statistical significance. Many books on statistical process control are available and most of them have control chart implementation as an integral part. Some of the books are largely based on statistical treatment, such as Banks (1989), Ryan (1989), Quesenberry (1997), Derman and Ross (1997), among others. Some are more practice oriented for quality engineers and managers; a selection of such books are Grant and Leavenworth (1996), Oakland (1986), Wetherill and Brown (1991), Wheeler and Chambers (1992), Besterfield (1997), Mitra (1998) and Montgomery (2001).

In this chapter, we discuss the need for process improvement and the problems that may be faced in controlling high quality processes. We first review some basic concepts in statistical process control with control charts, although it is assumed that the readers would have already been familiar with them. Section 1.1 reviews the notations and terminology to be used, most of which are in common with those in the literature.

1.1. Basics of statistical process control

Statistical process control originated in early twenties when Shewhart (1926) presented ideas of statistical control charts for process characteristics. The basic principle is that processes are always subject to random variation. Examples are the variation in the quality of incoming material, environmental factors such as temperature, humidity and vibration, and differences in product handling by different people (Figure 1.1).

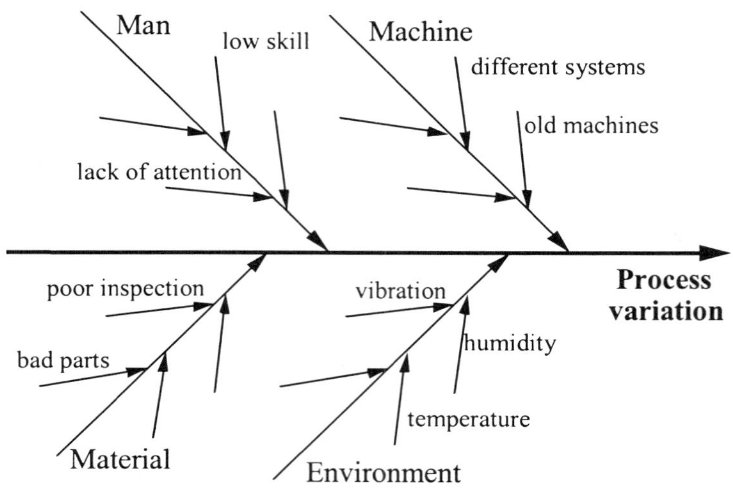

Figure 1.1. Common factors affecting process characteristics.

Random variation is generally not controllable or even identifiable; thus some variation within limits are allowed. In order to do this, we can plot successively observed process characteristics and make decisions based on what has actually been observed in the long run.

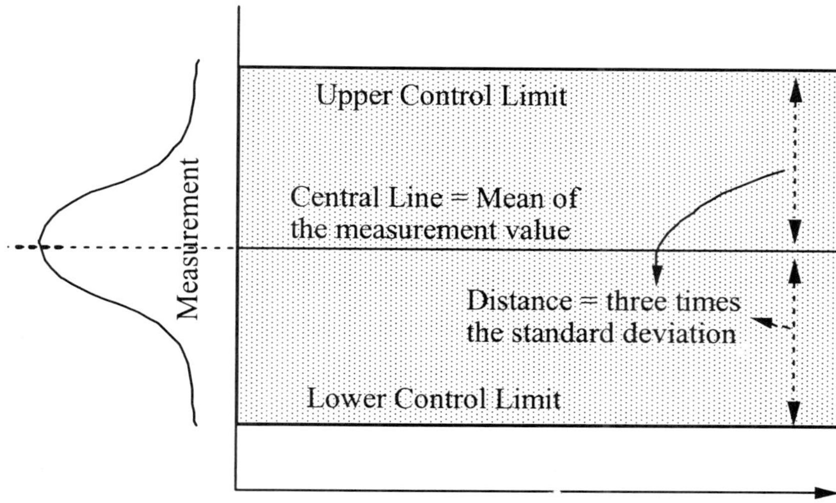

Figure 1.2. Basic principle of control chart with traditional 3-sigma limits.

Although there are many types of control charts and different ways to construct them, the common steps in setting up a control chart in practice can be summarized as follows.

(1) Obtain a series of process characteristics through observation, with or without calculations;

(2) Calculate the process mean and use it as the center line (*CL*);

(3) Calculate the standard deviation;

(4) Calculate the upper control limit (*UCL*) and the lower control limit (*LCL*) based on the mean plus and minus three standard deviation, or 3-sigma, respectively (Figure 1.2);

(5) Plot the process characteristics on the chart and connect the consecutive points;

(6) If there are points that fall outside of the limits, check the reason and decide on their disposal, then modify the *CL*, *UCL* and *LCL* if necessary;

(7) Continue plotting whenever a new measurement is obtained.

Control charts in industry are divided into two main types, variable charts and attribute charts. For variable charts, the process or quality characteristics take on continuous values while for attribute charts, the data is in the form of discrete counts. The information on product conformance to the required characteristics is usually given as count numbers or percentages. The p-chart is usually used to monitor the proportion nonconforming (sometimes described as defectives) in a sample. When we are interested in the number of nonconforming items in a sample of size n, the np chart can be used.

The control limits for the p- or np-chart can be derived in the following manner. It is well-known that the number of nonconforming items in a sample of size n follows the binomial distribution with parameter p. The mean of such numbers is np and the standard deviation is $np(1\text{-}p)$. Hence the control chart for p can be constructed with

$$\begin{aligned} UCL &= p + 3\sqrt{p(1-p)/n} \\ CL &= p \\ LCL &= p - 3\sqrt{p(1-p)/n} \end{aligned} \tag{1.1}$$

and for the np-chart, which is used for the monitoring of the number of nonconforming items in samples of size n,

$$\begin{aligned} UCL &= np + 3\sqrt{np(1-p)} \\ CL &= np \\ LCL &= np - 3\sqrt{np(1-p)} \end{aligned} \tag{1.2}$$

A typical p-chart is shown in Figure 1.3. When the sample size is fixed for all samples, the p-chart and np-chart are very similar and the difference is only in the scale of the y-axis. Different p-charts can be easily compared as the center line is the process fraction nonconforming level. For np-chart, the center line is affected by the sample size and when they are different, the levels are not directly comparable for different charts. However, the np-chart is a plot of a direct observation, i.e., the number of nonconforming items observed.

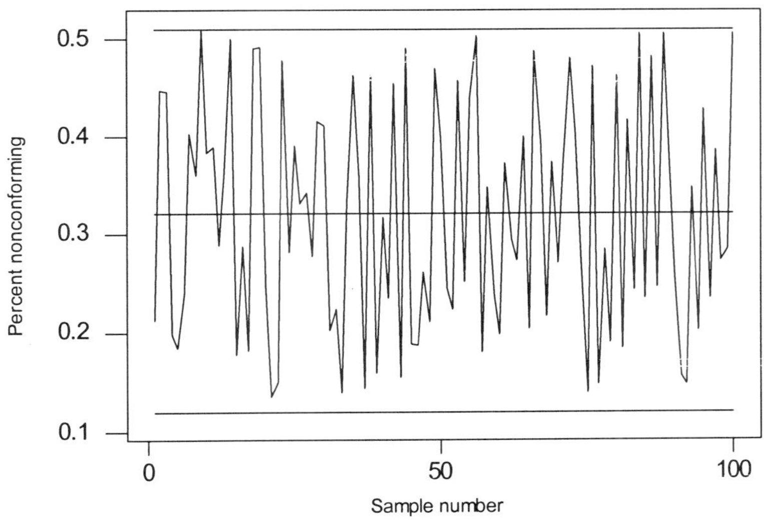

Figure 1.3. A typical p-chart with traditional 3-sigma control limits.

When the data is in form of the number of nonconformities (sometimes described as defects) in each sampled unit, the c-chart is commonly used. A related chart is the *u*-chart when the quantity of interest is the number of nonconformities per unit.

For the *c*-chart, the traditional approach is based on the Poisson model as it is well-known that the number of nonconformities in a sample follows Poisson distribution with a certain parameter *c*. The mean and variance for this distribution are given as *c*. Hence the elements of a *c*-chart are

$$\begin{aligned} UCL &= c + 3\sqrt{c} \\ CL &= c \\ LCL &= c - 3\sqrt{c} \end{aligned} \tag{1.3}$$

A typical *c*-chart is shown in Figure 1.4.

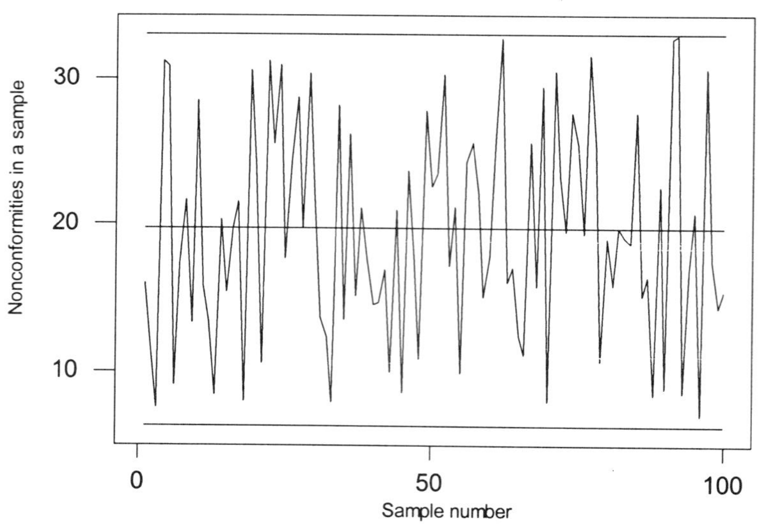

Figure 1.4. A typical *c*-chart with traditional 3-sigma control limits.

A u-chart is the plot of number of nonconformities per unit. The relationship between the c-chart and u-chart is similar to that between the p-chart and np-chart. That is, the scale is changed by a factor n, the sample size. The elements of a u-chart are

$$\begin{aligned} UCL &= u + 3\sqrt{u/n} \\ CL &= u \\ LCL &= u - 3\sqrt{u/n} \end{aligned} \tag{1.4}$$

When the process characteristics can take any value on a continuous scale, control charts for variables are used. The most common chart of this type is the $\overline{X}$ chart, which is usually associated with another chart for the process variation of the process, the R-chart. The principles of $\overline{X}$ & R charts are explained in standard works such as Grant and Leavenworth (1996). A typical $\overline{X}$ & R chart is shown in Figure 1.5.

It can be noted here that all the charts presented in this section are based on the normal approximation of the actual distribution of process characteristics. When 3-sigma control limits are used, the associated probability of false alarm, that is, the probability that a point falling outside the control limit when the process is in control, is about 0.0027. This has been regarded as a good level economically and commonly accepted in industry. The 3-sigma limits are also easy to compute and hence suitable for practical applications.

High-quality processes are usually associated with low counts of nonconforming items. For such a process, if accurate measurements can be taken to measure the severity of the nonconformity, variable charts could provide more information about the state of the process. On the other hand, such an approach usually requires the use of much more sophisticated measurement equipment. It may also require amount of time and manpower that may not be available or are costly. Hence our focus in this book is on the modification of the common attribute charts and the extension of the traditional charts without requirements for additional resource although in some later chapters, a number of issues related to variable charts are also discussed. Some related references on attribute charts are Woodall (1997) and Shore (2000).

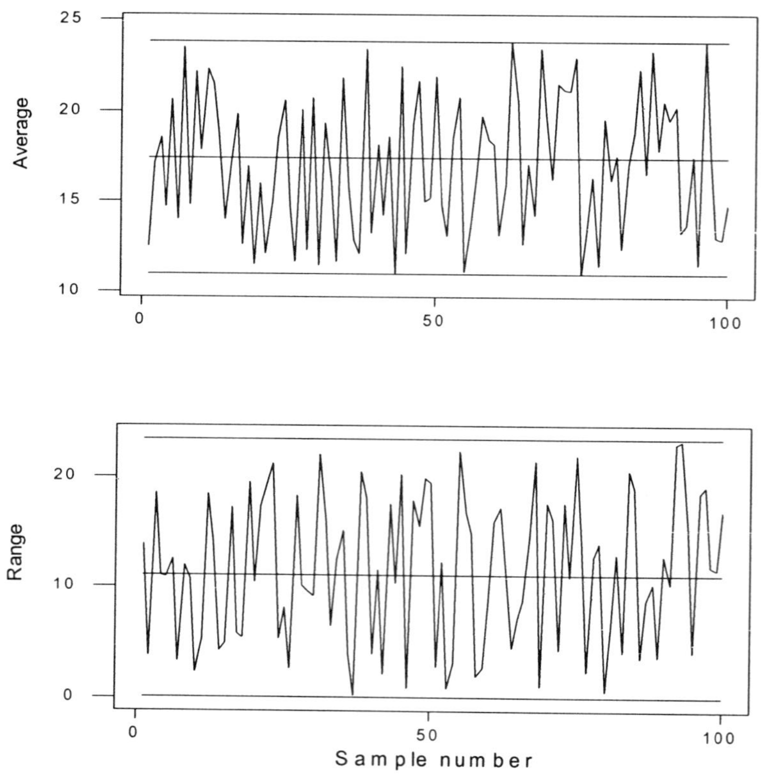

Figure 1.5. A X-bar and R- chart with traditional 3-sigma control limits.

1.2. Inadequacies of traditional attribute charts

Statistical process control based on Shewhart control charts, because of its simplicity in implementation and conciseness in interpretation of the state of a process, has been widely used in the manufacturing industry. However, because of the increasing effort of process improvement and rapidly developing technology, industrial processes have improved to a stage where many traditional control charts would face practical problems. The situation is more serious with attribute charts which, at the same time, is of increasing importance because of the possibility of obtaining count data quickly enabling processes to be monitored at a low cost.

We will first discuss some of the problems associated with attribute charts, mainly in terms of the *np*-chart and the *c*-chart.

1.2.1. Some problems associated with *p*-chart

Firstly, the traditional *p*-chart or *np*-chart are based on normal approximation. When p is small and the sample size is not large enough, the probability of false alarm is very high. Table 1.1 is generated by

$$P(False\ alarm) = 1 - P(LCL \leq X \leq UCL) \tag{1.5}$$

assuming that X follows binomial distribution with parameters n and p. The LCL and UCL are as in Eq.(1.1).

Secondly, the probability of failure in detecting a change of process characteristic will be high. Because the actual distribution of p is far from normal, the control limits based on the normal approximation can no longer describe the real state of the process; neither can it detect process changes accurately. A related problem is that, because of the problem of high false alarm probability, true signals may also be overlooked.

Table 1.1. Exact false alarm probability for *np*-chart with 3-sigma limits when *p* is small

p	n=5	n=10	n=20	n=50	n=100
.01	.9520	.9086	.8189	.6066	.3695
.02	.9078	.8179	.6747	.3674	.1367
.03	.8672	.7402	.5465	.2218	.0508
.04	.8160	.6710	.4494	.1335	.0191
.05	.7749	.5998	.3611	.0801	.0074
.06	.7359	.5406	.2957	.0480	.0047
.07	.6988	.4876	.2362	.0287	.0023
.08	.6636	.4402	.1925	.0171	.0026
.09	.6304	.3904	.1529	.0103	.0015
.10	.5910	.3503	.1240	.0084	.0023

Example 1.1:

To illustrate this, we use a case presented in Xie and Goh (1993). In Messina (1987), p.144, a control chart is given for average process proportion nonconforming p=0.0234. The conventional 3-sigma *UCL* and *LCL* are calculated as 0.0426 and 0.0043, respectively. If the *LCL* is used, then for a sample size of n=563 used in the original example, the process is out of control when two or less nonconforming items are found.

In that case, we can calculate the probability of the number of nonconforming items, X, as

$$P(X=0)=(1-p)^n=0.000002$$

$$P(X=1)=np(1-p)^{n-1}=0.000020$$

$$P(X=2)=\frac{n(n-1)}{2}p^2(1-p)^{n-2}=0.00015$$

Thus,

$$P(X\le 2)=0.000002+0.000020+0.00015\approx 0.00017$$

This is far from the assumed level 0.00135 for a traditional control chart with 3-sigma limits.

In this case, the lower limit should be set at least 3. The probability is then 0.00106, which is fairly close to the desirable level. This means that when there are only 3 nonconforming items in a sample (of size 563), the probability that the process is better than before is 1-0.001=99.9%.

□

Thirdly, in a near zero-defect environment, *LCL* will usually be less than zero, making it meaningless for a graphical plot. When counting the number of nonconforming items, process improvement cannot be detected unless some run rules are applied (Nelson, 1984). This will generally make the control chart more cumbersome to use. Even if it can be easily done by the computer, the common run rules will make the interpretation of the chart difficult to explain.

Last but not least, a more serious problem is that when the percent nonconforming is very small, even the *UCL* can be smaller than one unless the sample size is very large. In this case, every sample containing only one nonconforming item will cause a false alarm, which is not useful to the chart user at all and makes the control chart itself, especially the plotting, useless.

1.2.2. Some problems associated with *c*-chart

Similar problems are also associated with the *c*-chart or *u*-chart. If c is small, $c\le 0.09$, say, the upper control limit, Eq.(1.3), can be less than one

(Figure 1.6). Since one is the minimum number of nonconformity for an alarm, it is usually taken as the upper control limit.

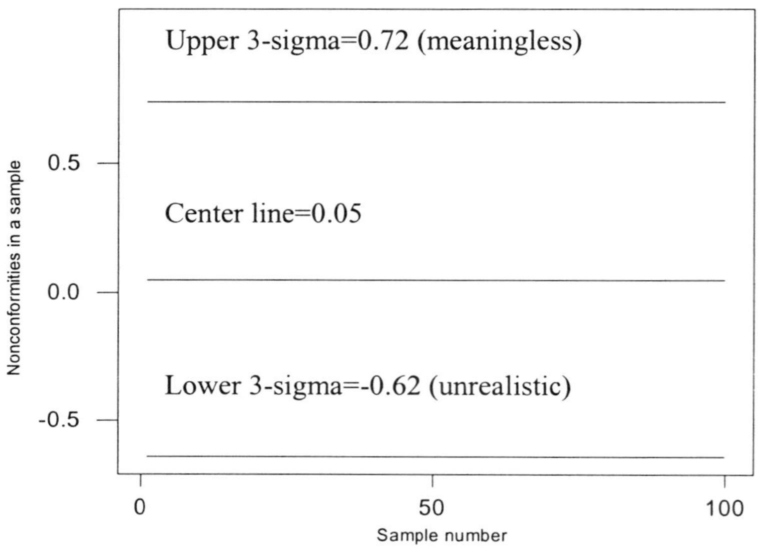

Figure 1.6. Unrealistic limits when the average process nonconformity is low; c=0.05 and control limits are computed with Eq.(1.3).

Note that since the number of nonconformities in a sample, C, is a Poisson distributed random variable, $P(C \geq 1)$ can be calculated as follows

$$P(C \geq 1) = 1 - P(C = 0) = 1 - e^{-c} \tag{1.6}$$

where c is the expected value of C, usually called the average number of nonconformities.

For small values of c, the approach will not be sensitive because of the low probability of occurrence of nonconformities. Also, even though it is thought conservative when we round the upper control limit up to one when the calculated value of UCL_C is lower in value, it can still give a much higher probability of false alarm than desired for a wide range of c because of the non-normality. For some values of c, Table 1.2 (Goh and Xie, 1995) shows the probability that one or more nonconformities will be detected.

Table 1.2. Actual probability of false alarm for some cases when the 3-sigma UCL_c has to be set to one

c	Prob($C\geq1$)
0.01	0.0100
0.02	0.0198
0.03	0.0296
0.04	0.0392
0.05	0.0488
0.06	0.0582
0.07	0.0676
0.08	0.0769
0.09	0.0861

In fact, even when $c>0.09$, the conventional UCL_C may not be suitable. Normal approximation to Poisson is good only when c is fairly large, say greater than 5. Unless the sample size is very large, for a high-quality process, the expected number of nonconformities is usually much less than this.

In Table 1.3 we calculate the probability of false alarm if the conventional UCL_C is used. Note that it can be more than ten times as large as the usual case. We also calculate the upper limit based on the exact out-of-control probability. The false alarm probability can be calculated in a similar manner using Eq.(1.5) assuming Poisson distribution, and the determination of the exact probability limits will be discussed in Chapter 2.

Table 1.3. Actual probability of false alarm and correct probability limit

c	3-σ UCL_c	P(false alarm)	Correct UCL_c
0.10	1.05	0.0047	3
0.15	1.31	0.0102	3
0.20	1.54	0.0175	3
0.25	1.75	0.0265	4
0.30	1.94	0.0369	4
0.35	2.12	0.0055	4
0.40	2.30	0.0079	4
0.45	2.46	0.0109	4
0.50	2.62	0.0144	5
0.55	2.77	0.0185	5
0.60	2.92	0.0231	5
0.70	3.21	0.0058	5
0.80	3.48	0.0091	6
0.90	3.75	0.0135	6

1.3. Need for quality improvement

Control charts are widely used in industry to detect significant deviations from process characteristics caused by assignable causes. It is commonly recommended that when an out-of-control signal is observed, the process be stopped to investigate such causes. This traditional view may not be in alignment with modern views which emphasize process improvement. Continues improvement is regarded today as the essence of survival and growth as the increasing competition in the global market has required ever-decreasing defective levels in processes.

Traditional control charts are used to maintain process performance, with their effectiveness lying in their ability to alert users of any deterioration of process behavior so that potential damage can be alleviated or averted in time. As a result, the attention of control chart users, mostly technical personnel, is mostly on performance deterioration. From a management point of view, the ability to detect process improvements in the form of "out-of-control" states is no less important, since improvements can take place as a result of process stabilization, sub-process improvement or learning effects.

Statistical process control can and should play an important role in endeavors for quality improvement. What is needed is a change of focus from prevention of process deterioration to detection of potential for process improvement, and to sustain the improvement effort. The techniques for process monitoring can be basically the same, although in near zero-defect environment, we have to pay attention to specific problems associated with the implementation of control chart technique and modify the approach used with the aim of allowing improvement to be detected. In this sense, a signal indicating a shift towards a non-random positive result, commonly called an alarm, might be a good welcome one from the point of view of process improvement.

Continuous improvement of a high-quality process, although difficult and not straightforward, is an important issue because of the need to meet increasing competition in the global market. Also, the high cost associated with products always implies a need for cost reduction. The consequence of failure in continuous improvement is deterioration in competitive position.

Example 1.2:
We have simulated a set of data illustrating the use of traditional chart in a high-quality process. A sample size of 200 is used and the fraction nonconforming level of 0.001 (which is 1000ppm). Most of the samples are conforming. Suppose that after the first 100 samples, the process is improved to a level of 0.0002 (200ppm). The sample size remains 200. An *np*-chart with the simulated data is shown in Figure 1.7. It is clear that this type of chart will not be of much use.

□

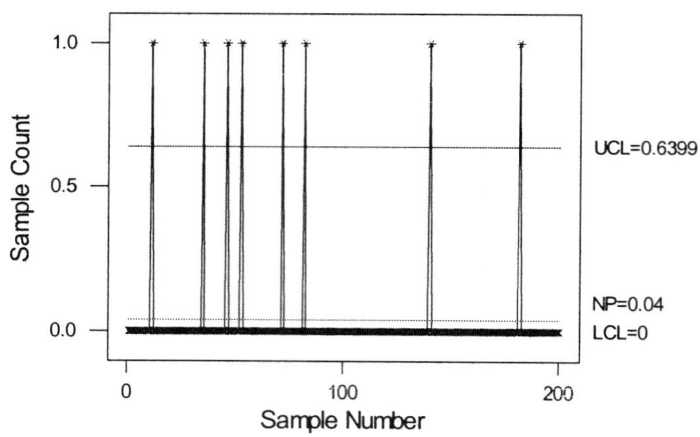

Figure 1.7. An *np*-chart for a high-quality process which was improved after sample 100.

In many basic statistical process control or SPC books, suggestion to use control charts for improved processes is made but the focus is usually on the calculation of new control limits. However, the problems mentioned in the previous section is a barrier for doing so when such process change cannot be detected by the chart in the first place. The failure of traditional control chart to detect process improvement, as well as the many practical problems associated with them when the defective rate is low, makes the subject of this book important in modern quality management.

1.4. Common probability distributions in SPC

In this section some commonly used probability distributions are described for reference in subsequent chapters. The distributions are not discussed in detail; however the descriptions and definitions are particularly useful in designing and analysis control charts to suit specific needs.

Bernoulli trials

If an experimental trial has two outcomes, success or failure, then it is known as Bernoulli trial. Let the probability of success at each trial be p and the probability of failure be $1-p$. The mean and variance of the distribution of the successes are, respectively, p and pq, with $q=1-p$. In the context of SPC, when an item can be conforming or nonconforming, the outcome of an inspection or test can be described as a Bernoulli trial.

Binomial distribution

Let X be a discrete random variable. If X denotes number of trials that result in an outcome of interest, then X has a binomial distribution with parameters n and p, where n is the number of trials. When a number of items are checked and the number of nonconforming items is counted, the number will follow a binomial distribution under the assumption that all items have the same probability of being nonconforming.

The binomial distribution with parameters n and p is defined as the distribution of a random variable X for which

$$B(x;n,p)=\binom{n}{x}p^x q^{n-x},\quad x=0,1,2,...,n \tag{1.7}$$

The mean and variance of binomial distribution are np and npq, respectively.

The value of p could be estimated when x_1, x_2, ..., x_n are a set of observed values from the binomial distribution. The estimator is given as

$$\hat{p}=\sum_i^n x_i / n \tag{1.8}$$

The binomial variable can be approximated by the normal variable with mean np and standard deviation $\sqrt{npq}$. This approximation is good when $npq > 5$ and $0.1 \le p \le 0.9$, or when $npq > 25$.

Poisson distribution

Let X denote the number of successes in a unit of a time. Then X has a Poisson distribution with parameter c

$$P(x;c)=\frac{e^{-c}c^x}{x!},\quad x=0,1,2,... \tag{1.9}$$

The mean and variance of the distribution are both equal to c. The Poisson distribution can be approximated by the normal distribution when the size of c is very large. This is a limiting distribution of binomial distribution when n tends to infinity, and p tends to zero while np remains c.

When c is unknown, it can be estimated by

$$\hat{c}=\sum x_i / n \tag{1.10}$$

Negative binomial distribution

Let X denote the number of trials until r success in a sequence of independent Bernoulli trial. Then X has a negative binomial distribution (or Pascal distribution) with parameter p and r. The probability function is given by

$$P\{X = i\} = \binom{i-1}{r-1} p^r (1-p)^{i-r}, \; i = r, r+1, \ldots \qquad (1.11)$$

The mean and variance of a negative binomial random variable are r / p and $r(1\text{-}p)/p^2$, respectively. Note that another form for negative binomial distribution is to model $X\text{-}n$ as it takes value of all nonnegative integers.

Geometric distribution

A special case of r=1 in the negative binomial distribution becomes geometric distribution and its density function is

$$G(x; p) = p(1-p)^{x-1}, \quad x = 1,2,\ldots,n,\ldots \qquad (1.12)$$

For instance, given a sequence of independent Bernoulli trials while assuming the probability of success at each trial is p, then a geometric variable can be defined as the number of trials up to the first success. The mean and variance of the geometric distribution are $1/p$ and $(1-p)/p^2$, respectively. The negative binomial random variable can also be seen as the sum of r independent identical geometric random variables with parameters $(1, p)$.

Exponential distribution

Let X be a continuous random variable. The exponential distribution with parameter λ is given by

$$F(x) = e^{-\lambda x}, \quad x \geq 0 \qquad (1.13)$$

The mean and standard deviation of the exponential distribution are the same and equal to λ. Exponential distribution is useful in modeling time between events when the time is a continuous random variable. In fact, geometric distribution is the discrete counterpart of exponential distribution.

Normal distribution

Let X be a continuous random variable. The normal distribution has two parameters, μ and σ. The probability density function is

$$f(x) = \frac{1}{\sqrt{2\pi}\sigma} \exp\left\{ \frac{-(x-\mu)^2}{2\sigma^2} \right\}, \quad -\infty < x < \infty \tag{1.14}$$

The mean and standard deviation of the distribution are μ and σ, respectively. Using the transformation $Z = (X - \mu)/\sigma$, the normal distribution becomes a standard normal distribution. The cumulative distribution function of a standard normal variable is denoted as $\phi(z) = P(Z \leq z)$.

Some simple relationships

The following relationships are well-known and useful.

1. The binomial variable to be approximated by the Poisson variable with mean np provided $p<0.1$ and $np<10$.
2. The binomial and negative binomial variables are related by $F_{NB}(n\text{-}x;x,p)=1\text{-}F_B(x\text{-}1;n,p)$.
3. The sum of r independent geometric variables is the negative binomial variable.
4. The Poisson variable is the limiting form of binomial variable as n tends to infinity, and p tends to zero such that np tends to c.
5. The geometric distribution is a discrete counterpart of exponential distribution and when x in Eq.(1.13) takes only integer values, the distribution becomes geometric distribution.

For further details, refer to Johnson and Kotz (1969), among others.

CHAPTER 2

Control Charts with Probability Limits

When the process characteristic does not follow the normal distribution, the probability limits should be used instead of the traditional 3-sigma limits. In a near zero-defect environment, many problems indicated in Section 1.2 will arise and the traditional control charts with 3-sigma limits cannot be interpreted in the same way as the actual false alarm probability could be very different. Probability limits can be obtained by fixing the false alarm probability at an acceptable level and is hence more appropriate. It is an easy task to compute the exact probability limits when the underlying distribution for process characteristics is known. In this chapter we focus on the need for and uses of probability limits in a high quality environment and for process improvement identification in general. Although the methodology applies to both variable charts as well as attribute charts, we will focus on attribute charts as the information is easily obtained and the problem of non-normality is more serious for attribute data.

2.1. Use of probability limits

Recall that when the traditional 3-sigma limits are used, the probability of false alarm is about 0.27%. If control chart users are to have consistent levels of confidence in their decision making, then both the *UCL* and the *LCL* should be obtained in such a way that this usual confidence level associated with Shewhart's 3-sigma concept is maintained. A process is judged out of control when the process, with a certain probability, performs worse than expected; in such a case actions should be taken to make the necessary corrections to the process. The process may also be "out-of-control" in the sense that it has become better than expected; in such a situation useful information can be sought about how the newfound improved performance may be sustained.

It is well known that the number of conforming items in a sample, or alternatively the number of nonconformities can often be modeled by binomial or Poisson distribution. These distributions are approximated by a normal distribution when the traditional 3-sigma control limits are used. However, there are problems if the 3-sigma limits are used mechanically without clear justification, especially when the sample size is small. We highlight some problems here and discuss what can be done with the use of probability limits. This issue is mentioned in some standard books, such as Wetherill and Brown (1991). Some more in-depth discussions and examples can also be found Xie and Goh (1993).

In fact, in a near ZD environment, the normal approximation is usually not valid and probability limits should be used to reduce the probability of false alarms. This was referred to in Section 1.2 in which some numerical values of the false alarm probability for binomial and Poisson distribution are given. For moderate level of fraction nonconforming, the use of probability limits can easily solve the problem by setting the appropriate false alarm probability accordingly.

Furthermore, in order to have a positive *LCL* for the *p*-chart when *k*-sigma limits are used, the following inequality has to be made valid

$$np > k\sqrt{npq} \tag{2.1}$$

which can be simplified as

$$p > \frac{1}{1+\dfrac{n}{k^2}} = \frac{k^2}{k^2+n} \tag{2.2}$$

If Eq.(2.1) does not hold for sample size *n* and fraction nonconforming level *p*, the *LCL* will be negative and no process improvement can be detected.

Example 2.1:
Suppose k=3 and n=50, then the *p* value has to be greater than 0.15 for the *LCL* to be positive. Note that this implies that the process is of very low quality. Hence, for a reasonably high-quality process, a simple *p*- or *np*-chart will not be able to detect process improvement unless some run rules (Nelson, 1984) are used.

□

When a *p*-chart is used for moderate or low fraction nonconforming levels, the only course of action for a positive *LCL* is to increase the sample size. However, this may not be possible in many situations.

For a *c*-chart or a chart for the number of nonconformities, the count may not follow the Poisson distribution itself in a near ZD environment. One such example is when the process is nearly ZD and certain uncontrollable random mechanisms may affect the product so that the number of nonconformities becomes large. In Section 6.2 we take a close look at this problem, which is common in the electronics industry. A modified Poisson distribution is proposed to model the measurement data and the probability limits are certainly more appropriate.

Another example is when the geometric distribution is applied to measurement data, which does not follow Poisson distribution. It is known that for geometric distribution, the *LCL* will nearly always be less than zero making it useless in practice (Xie and Goh, 1997). The probability limits have to be used to resolve this problem. The use of geometric chart is an important focus in this book and will be discussed extensively in Chapter 3.

2.2. Setting up a control chart based on probability limits

The setting up of a control chart based on probability limits is straightforward provided that the underlying distribution of the quality characteristics is known or accurately estimated. The difference between the use of probability limits and the standard control chart limits is that we use the underlying distribution and the probability limits instead of the more easily calculated 3-sigma limits.

Before constructing a control chart based on probability limits, we have to fix an acceptable false alarm probability level, which is denoted by α here. For the 3-sigma control limits under normal distribution, this is equal to 0.0027, although any other value can be used and this should depend on practical situations.

Given the value of α, the steps for setting up a control chart based on probability limits is the same as that of Section 1.1, except that Step 4 should be replace with the following:

(4*) Calculate the upper control limit (*UCL*) and the lower control limit (*LCL*) as the upper $1-\alpha/2$ and lower $\alpha/2$ percentile, respectively;

To derive probability limits the underlying distribution of the process measurement should be used. As illustrated in Figure 2.1, suppose that Z is the measurement value to be plotted, be it an average or the count in a sample. Let $F(Z)$ be the probability cumulative function of Z. The lower probability limit, z_l, and upper probability limit, z_u, can be obtained as the solution, or the nearest solution, to the equations

$$P(Z \le z_l) = F(z_l) = \alpha / 2 \tag{2.3}$$

and

$$P(Z \le z_u) = F(z_u) = 1 - \alpha / 2 \tag{2.4}$$

where α is a prescribed acceptable false alarm probability.

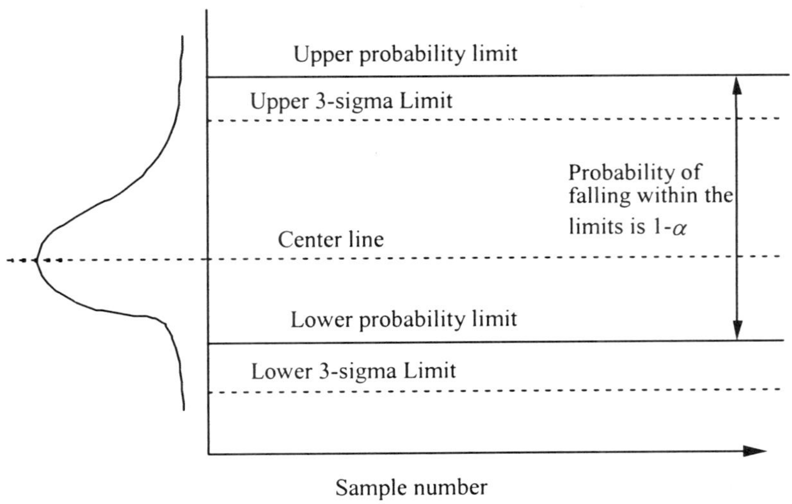

Figure 2.1. The probability limits should be used when the distribution of the process characteristics is not normal.

2.2.1. Lower limits for binomial and Poisson models

In this section, we study some common attribute charts, which are based on binomial and Poisson distributions. As the problems associated with 3-sigma limits are more serious for the lower control limit, we will focus on lower probability limits in this section, although the calculation of the upper probability limits are similar. The exact lower probability limits are presented analytically and numerically for some selected values. We first look at the chart for nonconforming items, or *np*-chart. The case with the fraction nonconforming chart, or *p*-chart, is similar.

When the *np*-chart is used we can calculate the exact probability that the number of nonconforming items *Z*, is less than, or more than, a given number. The control limits, especially the lower one, should also be set according to these probabilities. For simplicity we assume that all samples are of equal size since the generalization to variable sample size is obvious.

The probability that there are exactly k nonconforming items in a sample of size n is

$$P(Z = k) = \binom{n}{k} p^k (1-p)^{n-k},\ k = 0,1,\ldots,n \tag{2.5}$$

Usually we are interested in the probability that there are k or less nonconforming items in a sample of size n. This can be calculated by summing up the previous expression as

$$P(Z \le k) = \sum_{i=0}^{k} \binom{n}{i} p^i (1-p)^{n-i}, k = 0,1,\ldots,n \tag{2.6}$$

Thus for a 99.865% confidence upper limit, which is equal to the level for 3-sigma limits when the normal approximation is valid, we can determine the *UCL* by using the following relationship

$$99.865\% = \sum_{i=0}^{k} \binom{n}{i} p^i (1-p)^{n-i} \tag{2.7}$$

If a 99.865% confidence lower limit is required, we can determine a k which satisfies

$$0.135\% = \sum_{i=0}^{k} \binom{n}{i} p^i (1-p)^{n-i} \tag{2.8}$$

There are statistical tables for calculating this, especially for small n and moderate p values. As binomial distribution is very widely used in studies of sampling inspection, this calculation is not unfamiliar to quality engineers with some statistical proficiency. Statistical software or even some spreadsheet software can also be used for this purpose.

It should be noted that most binomial tables show the binomial probability for small *n* values only. For some combinations *n* and *p*, in Table 2.1 (Xie and Goh, 1993), numerical values of *k* to be used as the *LCL* are shown for some combinations of *n* and *p*.

Table 2.1. Some correct probability *LCL* for *np*-chart

n	Percent nonconforming, *p*						
	0.1	0.08	0.06	0.05	0.04	0.02	0.01
100	1	0	*	*	*	*	*
150	4	2	0	*	*	*	*
200	7	5	2	1	0	*	*
250	11	7	4	3	1	*	*
300	15	10	6	4	2	*	*
350	18	13	8	6	3	0	*
400	+	16	10	7	5	0	*
450	+	+	12	9	6	1	*
500	+	+	14	11	7	1	*
550	+	+	17	12	9	1	*
600	+	+	+	14	11	2	*
650	+	+	+	16	11	3	*
700	+	+	+	18	13	3	0
750	+	+	+	+	14	4	0
800	+	+	+	+	16	5	0
850	+	+	+	+	17	5	1
900	+	+	+	+	18	6	1
950	+	+	+	+	+	6	1
1000	+	+	+	+	+	7	1

"*": No solution - improvement detection not possible.
"+" *np* large enough (>25) to use conventional 3-sigma control limit.

It should be pointed out that as the binomial distribution is a discrete distribution, It is not possible to always meet the specified level of confidence exactly. For example, if a 99.865% confidence *LCL* is required for the case of n=200 and p=0.05, since k can only take on integer values, we have to choose between k=1 which gives a confidence level of 1-0.00040 and k=2 which gives a confidence level of 1-0.00234. Both are still quite far from the standard 1-0.00135 for which 0.00135 is known as the one-sided false alarm probability associated with 3-sigma limits.

The values *LCL* in Table 2.1 are determined by

$$\min\left\{k : \sum_{i=0}^{k} \binom{n}{i} p^i (1-p)^{n-i} \le 0.00135\right\} \tag{2.9}$$

If np is sufficiently large, we can use the conventional control limits since in that case, normal approximation of the binomial distribution will be good enough. On the other hand, if the sample size is very small and the proportion nonconforming is very low, we cannot even use any control chart based on one sample. Techniques of handling this situation by means of cumulative control charts will be discussed in Chapter 3.

For the c-chart, the problem is similar. Here we consider the upper control limit. The number of nonconformities needed to judge out of control status of a process at level α, based on one-sided upper limit can be calculated using

$$P(k \text{ or more nonconformities in the sample}) < \alpha\,. \tag{2.10}$$

That is, since the number of nonconformities in the sample is Poisson distributed with parameter c, the *UCL* for c-chart can be computed with the smallest k satisfying the following equality

$$\sum_{i=0}^{k-1} \frac{e^{-c} c^{i}}{i!} > 1 - \alpha \qquad (2.11)$$

The above inequality can easily be solved because k is usually a small integer. Some numerical values are tabulated in Table 2.2. Note that the interpretation is that when a value falls below the value listed, the process is likely to have improved, and causes can be found and sustained.

Table 2.2. Correct probability *UCL* based on Poisson distribution

c	Confidence Level α				
	.00135	.0027	.005	.01	.05
0.1	3	3	2	2	2
0.2	3	3	3	3	2
0.5	4	3	3	3	2
1.0	5	5	4	4	3
1.5	6	6	5	5	4
2.0	7	7	6	6	5
2.5	8	8	7	7	5
3.0	9	9	8	8	6
4.0	11	11	10	9	8
5.0	13	12	12	11	9

2.2.2. A note on some modified limits

When *p*- or *np*-charts are used, it is important to use the appropriate limits. Because of the skewness of the binomial distribution, the lower limit based on the 3-sigma concept may not exist as it is usually a negative value. Probability limits should be used when possible. On the other hand, simple modification can be used to obtain better control limits for *p*- or *np*-charts.

This section provides some improvements made on traditional *p*- and *np*-charts using some transformations so as to achieve high power in detecting upwards shift in a process. With simple adjustments to the control limits of the *p*-chart, one can achieve equal or even better results, while still working with the original scale. Some of the available methods are as follows.

(a) Binomial Q chart

By transforming the binomial variable to a standard normal variable, a Q chart (Quesenberry, 1997) could be drawn, with upper and lower control limits equal to 3 and –3 respectively. It is defined by

$$u_i = \sum_{k=0}^{i} \binom{n}{k} p^k (1-p)^{n-k} \tag{2.12}$$

and

$$Q_i = \phi^{-1}(u_i) \tag{2.13}$$

where $\phi^{-1}(\cdot)$ is the inverse function of the standard normal variable. Plotting Q_i's on the chart, a standard normal control chart with a uniform limit of 0 and 1 could be maintained.

(b) Arcsine p-chart

This is another nonlinear transformation like in the binomial Q chart, with

$$w_i = 2\sqrt{n_i}\left[\sin^{-1}\left(\sqrt{\frac{x_i + \frac{3}{8}}{n_i + \frac{3}{8}}}\right) - \sin^{-1}\left(\sqrt{p}\right)\right] \tag{2.14}$$

Since w_i is approximately standard normal, a chart could be drawn by plotting w_i on the chart with control limits are of 3 and –3.

In the case of unknown p values, the p value must be estimated first and then w_i can be used with p equal to $\hat{p}_{i-1}$.

(c) Cornish-Fisher expansion method

This is unlike the above two methods in that it is a direct method having some additional constant values in the three sigma limits using the normal distribution.

Define $y_i = x_i / n$. Plotting y_i values on the chart it is easy to detect the changes in p values with the following control limits.

$$UCL = p + 3\sqrt{\frac{p(1-p)}{n}} + \frac{4(1-2p)}{3n} \qquad (2.15)$$

and

$$LCL = p - 3\sqrt{\frac{p(1-p)}{n}} + \frac{4(1-2p)}{3n} \qquad (2.16)$$

When p is unknown, they have to be estimated and inserted in the above equations. It is also noted that the above equations could be used for variable sample sizes replacing $n = n_i$.

(d) Modified limits for np-chart

Ryan and Schwertman (1997) proposed the following regression equations of np and $\sqrt{np}$ and found them accurate. The proposed control limits are

$$UCL = 0.6195 + 1.00523np + 2.983\sqrt{np} \qquad (2.17)$$

and

$$LCL = 2.9529 + 1.01956np - 3.2729\sqrt{np} \qquad (2.18)$$

It is noted that these limits are valid for $p < 0.03$ while assuming the acceptable average run length (*ARL*), which is the average number of points plotted before an alarm, to be 370.4. In comparing with the exact limits, these limits are less biased in terms of *ARL* than the Shewhart *p*-chart.

Some other modifications can be found in Acosta-Mejia (1999), Shore (2000) and Kanji and Arif (2001). Ryan (1989) and Winterbottom (1993) also discussed some transformations for attribute chart to deal with the non-normality of discrete distributions. When the *p*- or *np*-chart is used, it is advisable to use the exact probability limits. The modified charts here are only for the purpose of improving 3-sigma limits in a simple way.

It can be noted that sometimes it is useful to have symmetric limits. The *Q*-transformation which is based on the exact probability distribution is the exact transformation in theory. However, because of the discrete nature of count data, the interpretation may not be as expected as that of a continuous distribution.

2.3. Probability limits for improvement detection

It is common to use control chart to monitor a process and judge it out of control when the process performs worse than expected. Actions are to be taken to make necessary correction to the process. However, the process may also be "out of control" in the sense that it has become better than average. This is a valuable opportunity not be missed, as continuous improvement is now the cornerstone of modern quality management. The improvement detection is usually associated with the *LCL* and when a point falls below the *LCL*, the process is supposed to have improved.

The most common problem with the 3-sigma *LCL* for attributes is that it is often less than zero. It is a common practice to ignore it or set it to be one. In the case of a Shewhart attribute chart, no improvement detection is possible. By using appropriate probability limits, this problem can be solved when the sample size is large enough.

In fact, when a zero-defect sample occurs, it may still tell us something about the process whether it is "out of control" in the sense that the process has become better than expected. It is highly desirable to be able to identify, with a known confidence, any improvement, which has occurred. The minimum size of a sample needed to identify a process improvement, based on a given known proportion nonconforming, can be established.

For a process with proportion nonconforming p, the probability that there is no nonconforming item in a sample of size n (n=1,2,...) is given by

$$P(\text{no nonconconforming in } n \text{ items}) = (1-p)^n \tag{2.19}$$

This expression, though simple, is very informative, since the probability can also be used to determine the minimum sample size needed to detect a process improvement over an existing proportion nonconforming p at a false alarm proability of α. The value of n can be expressed as

$$n = \frac{\ln \alpha}{\ln(1-p)} \tag{2.20}$$

Some numerical values of n calculated using Eq.(2.20) are presented in Table 2.3.

Note that this table gives us the probability that there are no nonconforming items in a sample of size n when the proportion nonconforming is p. However, we emphasize the lower part of the distribution here for the purpose of detecting process improvement.

Table 2.3. Sample size required identifying a process improvement at certain level α when process proportion nonconforming is p

p	α						
	.001	.00135	.0027	.005	.01	.02	.05
.005	1379	1319	1180	1057	919	781	598
.010	688	658	589	528	459	390	299
.015	458	438	392	351	307	259	199
.020	342	328	293	263	228	194	149
.025	273	261	234	210	184	155	119
.03	227	217	195	174	152	129	99
.04	170	162	145	130	116	96	74
.05	135	129	116	104	93	77	59
.06	112	107	96	86	75	64	49
.07	96	92	82	74	65	54	42
.08	83	80	71	64	56	47	36
.09	74	71	63	57	49	42	32
.10	66	63	57	51	44	37	29
.15	43	41	37	33	29	25	19
.20	31	30	27	24	21	18	14
.25	24	23	21	19	17	14	11
.30	20	19	17	15	13	11	9

Example 2.2:
To illustrate the idea, we use the number of nonconforming items in Table 2.4 for a given process average p=0.005 and sample size of 1319 with a false alarm probability of 0.00135. It is clear that the probability limits should be used in this case. Based on this, the limits are set at 0 and 17 for lower and upper controls.

Suppose that at the end, the fraction nonconforming level is changed to p=0.003, as the chart with probability limits suggests. With sample size of 1319, no improvement could be detected even with probability limits shown in Figure 2.2. If this is desirable, the sample size could be increased to a value higher than 1319. If there is no nonconforming item even in such a large sample, it is likely that the process has further improved.

□

Table 2.4. A set of nonconforming data from the Binomial distribution (read top-down and left-right)

p=0.005, n=1319				p=0.003, n=1319			
5	5	7	5	5	4	3	2
7	6	8	11	6	3	3	3
3	5	8	8	6	6	3	5
4	3	7	6	6	5	4	4
7	5	6	9	3	3	3	1
8	6	5	6	0	3	2	6
9	7	5	5	3	3	3	1
5	9	5	6	1	1	3	2
6	6	7	6	11	2	3	4
8	3	4	4	2	6	1	3
8	12	3	5	1	8	4	2
7	6	1	8	5	2	6	5
5	6	12	11	2	4	3	8
5	5	3	8	3	3	5	4
8	4	5	7	6	3	2	4
11	13	8	8	1	4	6	3
6	3	6	8	3	5	3	0
7	7	8	4	2	6	4	4
5	4	5	4	1	3	3	6
6	7	6	8	3	6	4	2
7	9	5	3	2	2	3	4
4	2	5	9	4	4	2	3
6	4	6	5	5	5	2	4
5	8	8	7	6	4	6	3
9	5	6	8	6	5	3	2

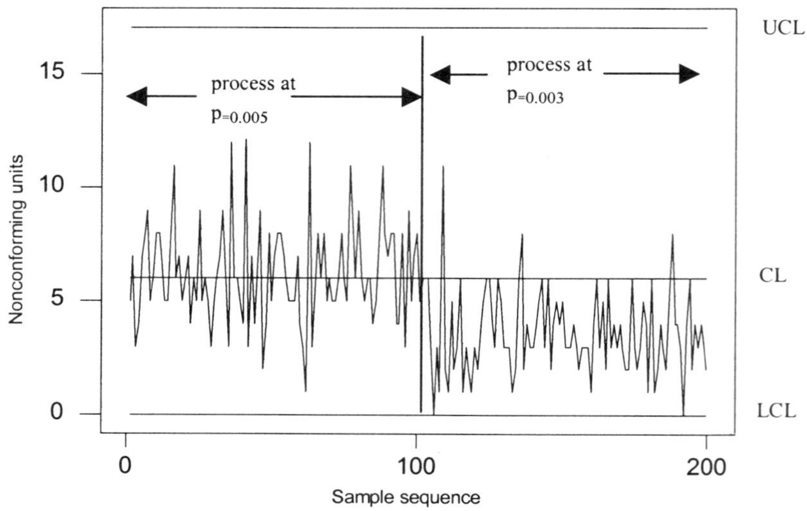

Figure 2.2. Probability limits for controlling process based on data in Table 2.4.

2.4. Limitations with probability limits

Although the advantages of using the probability limits are clear, there are some limitations. It is not a universal tool for process control and it is related to the specific underlying process characteristics we are measuring. Here we will discuss some potential problems with the traditional *np*-chart and *c*-chart even if probability limits are used. In fact, although the use of probability limits as described in Section 2.2 should be encouraged when implementing control charts in general, especially for non-normal distributed quantities, their usefulness for the *np*-chart and *c*-chart is limited.

For the *np*-chart and *c*-chart, when the process quality has improved, it is still very common that the lower control limit will be less than zero because the sample size cannot be increased to a very large value in practice. Hence, the problem of not being able to detect process improvement remains unless run rules are used, which would make the implementation and especially the interpretation more difficult. When the process is of very high quality, the upper control limit may also be less than one as in the case of 3-sigma limits. In that case, any nonconforming item or nonconformities will lead to the conclusion that the process is out of control. This makes the control chart useless from a monitoring point of view.

For traditional attribute charts, it is usually not possible, using the information from a single sample, to determine that the process has performed better than expected, when both the proportion of nonconforming items is very low and the size of the sample is small, i.e. smaller than that recommended by Table 2.3. Generalization of the idea used here is possible by looking at the number of consecutive zero-nonconforming samples such as techniques detailed in chapters 3 and 4, and some extensions in the following chapters.

Another practical problem is that the probability limits are usually not symmetric and engineers may want to have symmetric limits in order to check whether the plotting is properly done or the chart is well-constructed. This is a practical problem as the use of symmetric limits for asymmetric distribution is highly misleading. Sometimes a suitable transformation of data can be made so that the limits can be more or less symmetric.

The problems described above require a new paradigm in managing high-quality processes. To this end, the rest of this book will focus more on the idea of monitoring the number of conforming items between two nonconforming ones instead of the approach used in traditional control charts. However, the probability limits are assumed to be used in the development of the control charts to be discussed. Using probability limits enables us to provide the chart signals an appropriate interpretation, which is important in practical applications.

CHAPTER 3

Cumulative Count of Conforming (CCC) Chart

The *CCC* chart, or the count of conforming control chart, is first developed in Calvin (1983) to monitor zero-defects processes. It was further studied and given the name *CCC* chart in Goh (1987). This is a powerful technique for process control when a large number of consecutive conforming items are observed between two nonconforming ones. A control chart can be set up to monitor this number and decisions made based on whether this number is too large or too small. In this chapter, this type of control chart is described in detail. Decision making procedures are presented for various situations. Different versions of *CCC* chart are also discussed. The *CCC* chart is very useful for one-at-a-time inspections or tests which are common in automated manufacturing process. It is generally a technique for high quality processes when nonconforming items are rarely observed.

3.1. The basic idea

When dealing with near ZD processes, there are very few nonconforming items and we usually have a long series of conforming items before a nonconforming one is observed. The traditional Shewhart control chart encounters many practical difficulties, such as physically meaningless control limits, too many false alarms and failure in detecting process improvement, as described in Section 1.2. Other means have to be developed to handle these problems.

The idea behind the *CCC* chart is the fact that the number of conforming items between two consecutive nonconforming ones changes when the fraction nonconforming shifts. It is not easy to observe the shifts themselves in near ZD processes because they are usually very small in absolute magnitudes. A strategy is to observe the cumulative count of conforming items between two consecutive nonconforming ones. Note that unless specified, we count the last item that is nonconforming in the *CCC*, and this can be interpreted as the number of conforming items that are inspected until a nonconforming item is observed.

It is clear that if the *CCC* is getting smaller, the process is producing too many nonconforming items over time and can be judged out of control. On the other hand, if the number is getting larger, then the process has probably improved and this can be used to identify further improvement potential so that continuous improvement can be achieved even in a near ZD environment.

It is interesting to note that Chan *et al.* (2000) used the idea of counting the cumulative confirming items in the case of continuous production process. When the production quantity is measured with volume, weight, length etc., the cumulative quantity produced between the occurrences of two defects can be used to monitor the production process. Note that the cumulative quantity can take any value, not necessarily an integer as the case of count data.

In the following sections, we give an introduction to the use of the *CCC* chart and discuss some basic issues. Since the development of the idea in Calvin (1983) and the introduction of *CCC* concept by Goh (1987), the use of *CCC* type control chart has been further studied by Lucas (1989),

Bourke (1991), Xie and Goh (1992), Ermer (1995), Glushkovsky (1994), Xie *et al.* (1995), McCool and Joyner-Motley (1998), Wu *et al.* (1999, 2000), among others. In the later chapters, several extensions and modifications are also discussed.

3.2. Setting up the *CCC* chart

The setting up of the *CCC* chart is similar to the generic procedure of setting up a Shewhart control chart, except that the measurement are the number of conforming items after the last nonconforming one. This count should be plotted only when a new nonconforming item is observed. The determination of the control limits will be presented in details here.

Let n be the number of items to be observed before a nonconforming one is found, i.e., n-1 items are conforming followed by the n:th item which is nonconforming. It is clear that this count follows a geometric distribution. Control limits can then be easily obtained based on this.

Mathematically, if the probability of a nonconforming item is equal to p, then the probability of getting (n-1) conforming ones following a nonconforming one is

$$p(1-p)^{n-1},\ n=1,2,... \tag{3.1}$$

The mean of the geometric distribution with parameter p, which can be used as the center line, is

$$CL = 1/p \tag{3.2}$$

Suppose that the acceptable of false alarm probability is α. The *UCL* and *LCL* can be determined as

$$UCL = \frac{\ln(\alpha/2)}{\ln(1-p)} \tag{3.3}$$

and

$$LCL = \frac{\ln(1-\alpha/2)}{\ln(1-p)} \tag{3.4}$$

respectively. Control charts can then be set up by including the control limits and the center line. Plotting can then be done in the usual way.

It is clear that the control limits are highly asymmetric. Hence, it is suggested that the log-scale be used for the plotting.

Example 3.1:
A set of *CCC* data is shown in Table 3.1. In Figure 3.1 the corresponding *CCC* chart is displayed. In this case, all points are within the control limits and the process is in control.

□

3.3. Decision-making related to *CCC* chart

Decision-making based on a *CCC* chart is straightforward. On the other hand, although the plot is similar to the conventional Shewhart charts, as we are measuring the number of consecutive conforming items, it is clear that when the chart shows an out-of-control signal in form of a value larger than the *UCL*, the process has probably improved. Process deterioration should be signaled at values that are lower than the *LCL* value. In that case, we have fewer consecutive conforming items between two nonconforming ones (Xie and Goh, 1992). Figure 3.2 presents a simple charting and decision making procedure for a *CCC* chart.

The *CCC* control chart has been devised much along the same line as the traditional Shewhart control charts. It assumes a fairly reliable knowledge of the steady-state proportion nonconforming p, and the control limits are associated with a predetermined α, i.e. the probability of judging a process out of control when it is not. Once the *CL*, *LCL*, and

UCL are determined, the user of the chart may not need to be aware of the underlying p and α values. Such a situation may be acceptable when p is fairly stable.

Table 3.1. A set of nonconforming data from a typical near ZD environment for p=500ppm

Defect sequence	*CCC*	Defect sequence	*CCC*	Defect sequence	*CCC*	Defect sequence	*CCC*
1	227	26	678	51	409	76	2196
2	2269	27	2088	52	4845	77	1494
3	1193	28	1720	53	4809	78	1906
4	4106	29	1656	54	504	79	548
5	154	30	201	55	257	80	987
6	12198	31	3705	56	702	81	6216
7	201	32	4042	57	4298	82	704
8	9612	33	716	58	1320	83	6477
9	4045	34	2010	59	1845	84	233
10	678	35	402	60	4641	85	855
11	2088	36	539	61	2815	86	188
12	1720	37	2665	62	903	87	4133
13	5562	38	1711	63	755	88	780
14	4042	39	1602	64	565	89	315
15	716	40	71	65	973	90	1425
16	2010	41	546	66	2555	91	580
17	402	42	655	67	1822	92	957
18	539	43	2065	68	4324	93	1443
19	8465	44	286	69	1140	94	3880
20	2269	45	1385	70	109	95	1357
21	1193	46	354	71	1981	96	234
22	4106	47	934	72	387	97	1836
23	154	48	3539	73	3268	98	7984
24	2011	49	1671	74	2666	99	110
25	4045	50	3955	75	5498	100	128

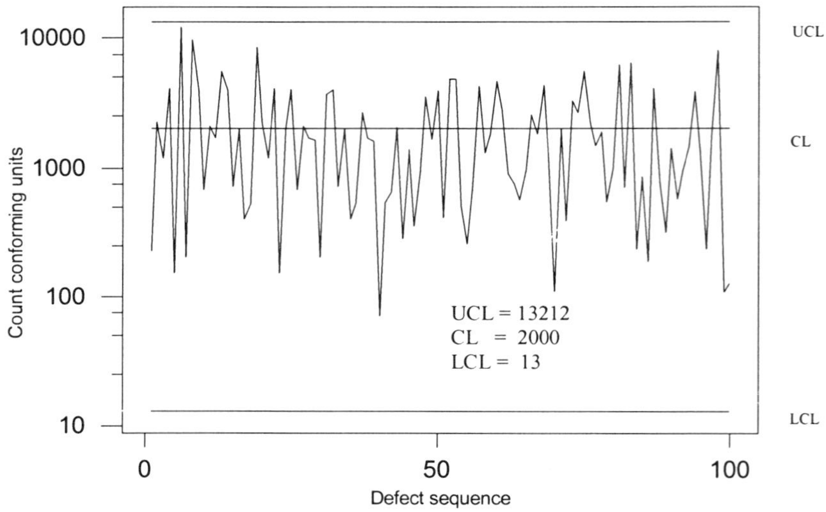

Figure 3.1. *CCC* chart for data in Table 3.1.

On the other hand, when p is extremely low, then its value would be more sensitive to short-term drifts in process parameters, and the process controller might want to focus more on the current p value in determining whether the process in question is out of control. For this purpose α need not be fixed, since it is in fact a confidence level parameter for the decision maker. Hence, a reference graph is developed in Xie and Goh (1992) to enable its user to judge the state of control (or otherwise) of a process via the p-α relationship.

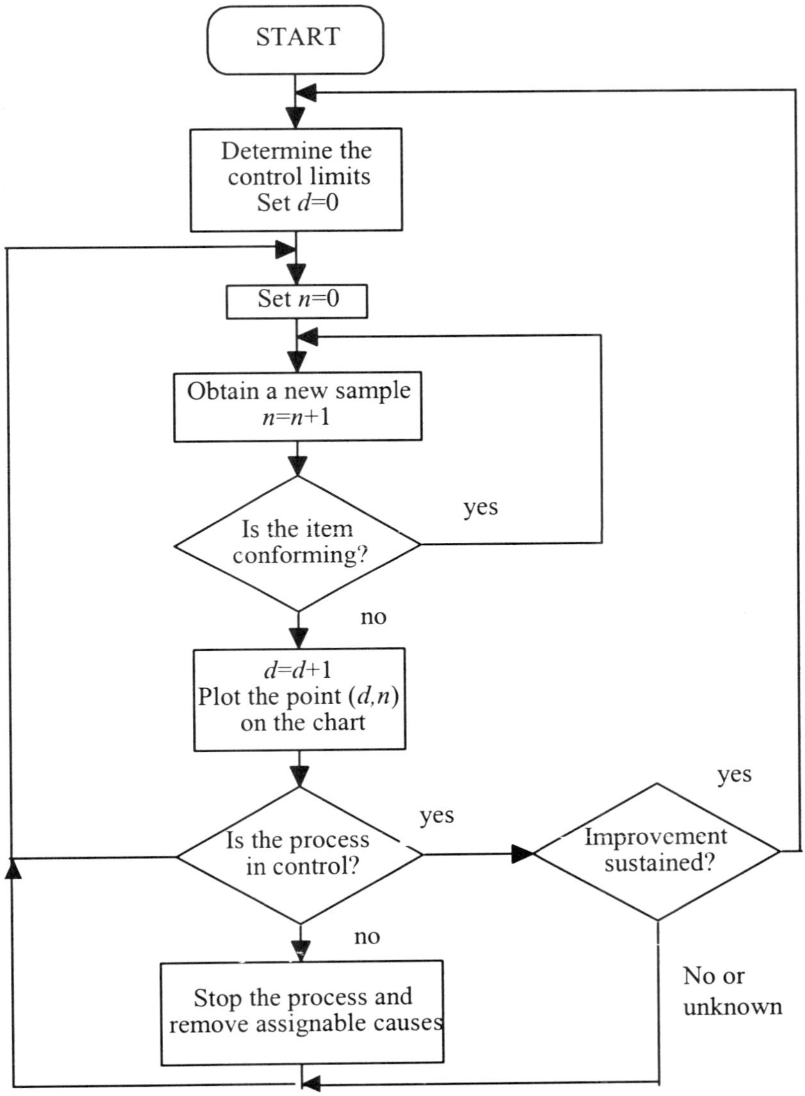

Figure 3.2. A charting procedure for *CCC* chart.

Another practical question arises when, during the application of the *CCC* control chart, a nonconforming item is found but the process is judged not out of control. According to the charting procedure, cumulative counts of conforming items should restart with zero even when the process is still in control. A procedure is now proposed to permit the use of gathered information up to this point, rather than ignoring such information and starting a fresh count.

It is possible, based on simple probability calculations, to establish the equivalent value of the current cumulative count for an in-control process that has just yielded one non-conforming item. This equivalent value is then used as the starting point of the next segment of cumulative counts, thus providing a more rational basis for the subsequent judgments on the state of the process. This will be discussed in detail in next section.

3.3.1. False alarm probability and control limits

Our discussion will start from a situation where a number of conforming items have been observed, and a single nonconforming item has just been found. It is required to decide whether the process is out of control, with an accompanying statement of the certainty level of this judgment. When using any control chart, false alarm is unavoidable because of the intrinsic process variability and sampling uncertainty.

It is understood that at the high yield level being considered, a process is presumed under control as long as there is no nonconforming item in a series of tests or inspections; the process is taken to be out of control when two or more nonconforming items are found. When one nonconforming item is found, the number of conforming items counted prior to it, n, is known. If the proportion of nonconforming items associated with the process is p, then the probability that no nonconforming item is found in n inspected is

$$P(\text{no nonconconforming in } n \text{ items}) = (1-p)^n \tag{3.5}$$

This probability reflects the certainty we have when we judge the process to be out of control when a nonconforming item has been found. Note that this is a decreasing function of n and p.

Xie and Goh (1992) introduced the concept of certainty level s, which is the probability that the process is actually out-of-control. The certainty level is related to false alarm probability when interpreting *CCC* chart signal. In the with one-sided control limit, $s=1-\alpha$. The relation between the proportion of nonconforming items p and number of items inspected n, which is given as

$$(1-p)^n = s \tag{3.6}$$

We can now determine the number of conforming items inspected before a nonconforming one is allowed for the process to still be considered in control. This can be calculated as

$$n = \frac{\ln s}{\ln(1-p)} \tag{3.7}$$

In Table 3.2 we give some numerical values of n for some different combinations of values of p and s. If the cumulated count of conforming items is less than the tabulated value, then the proportion of nonconforming items is higher than p with certainty s. This table can be easily expanded for other values of p and s.

For practical applications, a *decision graph* can be constructed to facilitate decisions on the state of control of a process whenever a nonconforming item is observed. A typical decision graph is given in Figure 3.3.

Table 3.2. If the number of conforming items is less than the tabulated value, then the proportion of nonconforming items is higher than p with certainty s

p	s			
	0.90	0.95	0.98	0.99
.0001	1053	512	202	100
.0002	526	256	101	50
.0003	351	170	67	33
.0004	263	128	50	25
.0005	210	102	40	20
.0006	175	85	33	16
.0008	131	64	25	12
.0010	105	51	20	10
.0015	70	34	13	6
.0020	52	25	10	5
.0030	35	17	6	3

The decision graph can be used in a number of ways. For example, we can, for any cumulative count n prior to a nonconforming item,

(1) for a given level of proportion nonconforming p, read the certainty level s from the graph; or

(2) for a given level of certainty s, read the value of p and see if it is higher than the acceptable level for the process; or

(3) for given p and s, determine the minimum number of conforming items that must have been observed before a nonconforming one can be tolerated, i.e. the process can still be deemed under control even one nonconforming item has been observed.

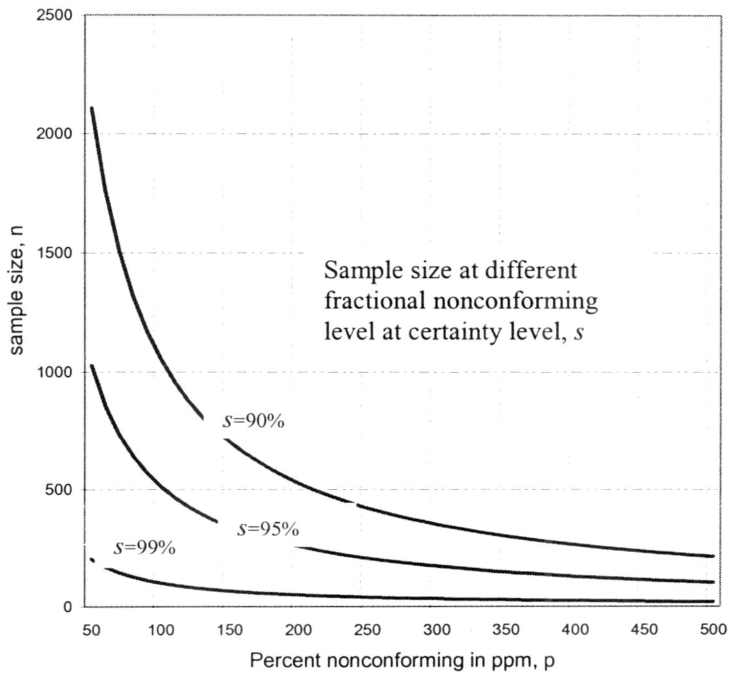

Figure 3.3. A decision graph for the state of control of a process.

Since usually $p<<1$ for high-quality products, a good approximation of Eq.(3.6) is

$$s = e^{-np}, p \leq 1 \tag{3.8}$$

Hence, for high-quality products, we have an exponential relation between the number of inspected items n, and the confidence that the process is out of control when a single nonconforming item is found.

The exponential relation is very convenient to use from another point of view. By taking logarithms, we can show that the relation is equivalent to

$$\ln n + \ln p = \ln[-\ln s] \tag{3.9}$$

Hence, the relation between $\ln n$ and $\ln p$ is linear for small values of p.

The procedure presented here goes beyond answering the question as to whether the process p is still at its original value (i.e. whether the process is in control); it allows the user to estimate the actual p with a chosen s. In high quality (i.e. low p) processes it is often important to track p for quality vigils and our procedure suits the purpose extremely well.

It is useful to note that the procedure can be used regardless of whether there is 100% inspection. Also, sample size does not matter here and the only quantity we have to update is the total number of consecutive conforming items inspected. Thus, for example, if k samples, each containing n_i items, have been inspected before a nonconforming item is found, then n is taken as

$$n = \sum_{i=1}^{k} n_i \tag{3.10}$$

3.4. *CCC* chart for grouped measurement

In using the cumulative count procedure, each time when a nonconforming item is encountered, we can judge whether the process is out of control. If it is, we have to rectify the process and start from the beginning by setting the count to zero because the process characteristics have changed. However, it may not be quite reasonable that one should start the count from zero again when the process is judged not out of control. We could utilize the hitherto gathered information about the process as far as possible. In this section, we study the problem of determining a reasonable starting point when a nonconforming item has been found and the process is not judged out of control.

Suppose that n_c items have been inspected and a nonconforming one is found. Suppose also that n_c is large enough not to lead to the judgment that the process is out of control. We may want to have the count set to another number n_0 depending on the current count of consecutive conforming items since, in this case, the process is not yet out of control and all the information about the process can still be useful.

A method is presented in Xie and Goh (1993) for the determination of n_0 to which the cumulative count should be set. It is obvious that when n_c is large, then n_0 should be large and vice versa, but the problem is to determine the appropriate value of n_0, as depicted in Figure 3.4.

We can relate the information that there is one nonconforming item in n_c to n_0 conforming ones by considering the probability that we will get two or more nonconforming items in n_c inspected items. Let this probability be equal to the probability that we will get one or more nonconforming items in n_0 inspected. We have the following equality

$$1-(1-p)^{n_0} = 1-(1-p)^{n_c} - n_c p(1-p)^{n_c-1} \tag{3.11}$$

The left-hand side above is probability that there are one or more nonconforming items in n_0 items and the right-hand side is the probability that there are more than one nonconforming items in n_c items inspected.

The equality for the determination of n_0, Eq.(3.11), can be rewritten as

$$(1-p)^{n_0} = (1-p)^{n_c} + n_c p(1-p)^{n_c-1} \tag{3.12}$$

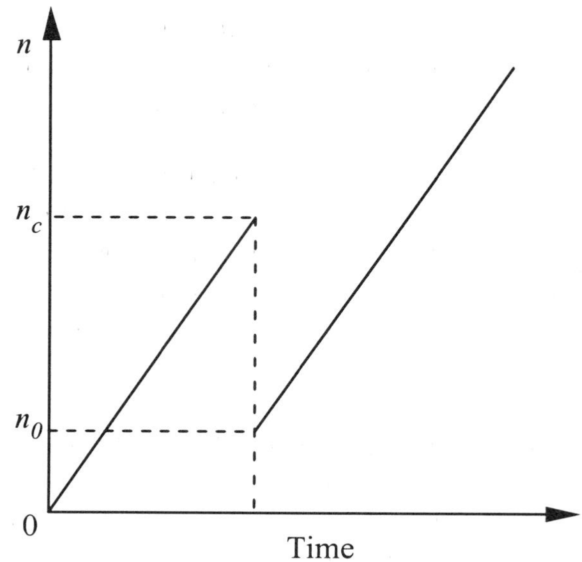

Figure 3.4. Use of an equivalent cumulative count of conforming items when one nonconforming item is found after n_c conforming ones.

By taking natural logarithm on both sides of Eq.(3.12), we get the solution of n_0 as

$$n_0 = \frac{\ln\left\{(1-p)^{n_c-1}(1-p+n_c p)\right\}}{\ln(1-p)} \tag{3.13}$$

It is interesting to note (Xie and Goh, 1993) that for small value of p, which is the case here, a good approximation of n_0 may be obtained based on the fact that

$$\ln(1-p) \approx -p - \frac{p^2}{2};\ p << 1 \tag{3.14}$$

Thus for $p<<1$, we have the following approximation

$$n_0 = \frac{(n_c - 1)\frac{p^2}{2} + (n_c - 1)^2 \frac{p^2}{2}}{p} = n_c(n_c - 1)\frac{p}{2} \qquad (3.15)$$

Some numerical values of the approximation are tabulated in Table 3.3. Since for high-quality products, $n_c p < 1$ so that n_0 is always less than $n_c/2$, which is expected. However, if $n_0 p$ is much less than 1, n_0 can be much less than n_c, in which case n_0 can usually be neglected and resetting is not necessary.

To summarize the procedure: when a nonconforming item is detected, the process has to be judged whether out of control. If it is, the next cumulative count will start from zero after the necessary process rectification. If it is not, then the cumulative count will resume with the value n_0, instead of n_c or zero. The procedure is summarized in the flow chart in Figure 3.5.

Table 3.3. Some numerical values of n_0

n	p				
	0.0001	0.0002	0.0003	0.0004	0.0005
100	0	1	1	2	2
150	1	2	3	4	5
200	2	4	6	8	10
250	3	6	9	12	15
300	4	9	13	15	22
350	6	12	18	24	30
400	8	16	24	32	40
450	10	20	30	40	50
500	12	25	37	50	62
600	18	36	54	72	90
700	24	49	73	98	122
800	32	64	96	144	160

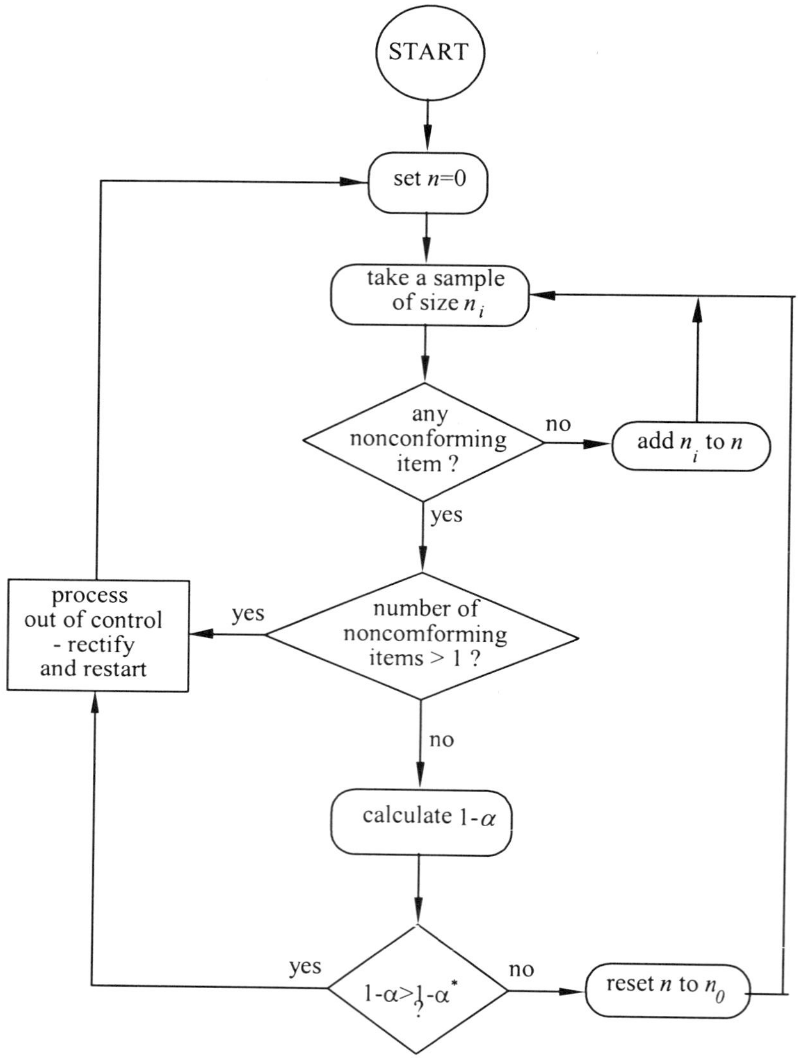

Figure 3.5. Statistical decision making based on cumulative counts of conforming items.

Example 3.2:
Using the data in Goh (1987), the use of the flowchart yields the result shown in Table 3.4.

Table 3.4. Application of the decision making flowchart to data of cumulative counts used in Goh (1987)

Instance of nonconforming	*CCC*	Adjusted count n	Process out of control ?	Reset count to n_0
1	585	585	no	68
2	55	123	yes	0
3	8126	8126	no	4509
4	2017	6526	no	3317

It is noteworthy that the values of n_0 as shown in Table 3.4 for resetting the counts at instances 3 and 4 are very large; under such circumstances, if we immediately observe another nonconforming item, the process is not judged out of control. This can be simply explained that at that time the process is much better than expected.

□

The *CCC* chart presented in this chapter is slightly different from the original one introduced in Calvin (1983). The original chart has a few drawbacks. The points plotted do not have uniform meaning and some represent the count of conforming items until a nonconforming one is detected while others represent the count of conforming items when a sample is exhausted. Most of the points are actually not used in decision making. Although it is assumed that samples of size n are available, it is still assumed that the inspection is done one-by-one.

The starting and end points are actually random if it is only known that there is a nonconforming item in the sample. To demonstrate this, a chart is shown in Figure 3.6 using Calvin's method for which a set of first 16 points chosen. Assuming sample size as 1000 and the limits values are found using Calvin's method.

On the other hand, the continuous plotting method by Calvin allow an upward shift in *CCC* level be detected before a nonconforming item is observed. This is important from point of view of improvement detection. It can be questionable if the process is judged to have improved when a nonconforming item has just been observed. Instead, as usually is the case, the process could have been improved long before the occurrence of nonconforming item and this can be captured by Calvin's method. Hence, it is important to check for positive assignable causes whenever a count has just exceeded the *UCL* of the *CCC* chart, no matter whether a point is plotted or not.

It can be pointed out that another useful idea in the original work by Calvin is that the plotting should be done on a semi-logarithmic scale. This is because of the asymmetry of the control limits when the distribution is very skewed. A plot using a semi-logarithmic scale is shown in Figure 3.7 and compared with Figure 3.6, it has a more traditional layout as other Shewhart charts.

Besides the papers cited in this chapter, there are some other related papers presenting idea of counting consecutive conforming items for constructing a control chart. There are also numerous extensions that have been studied. Chapter 5 and Chapter 6 present some advanced cases related to the extensions of the model or implementation issues.

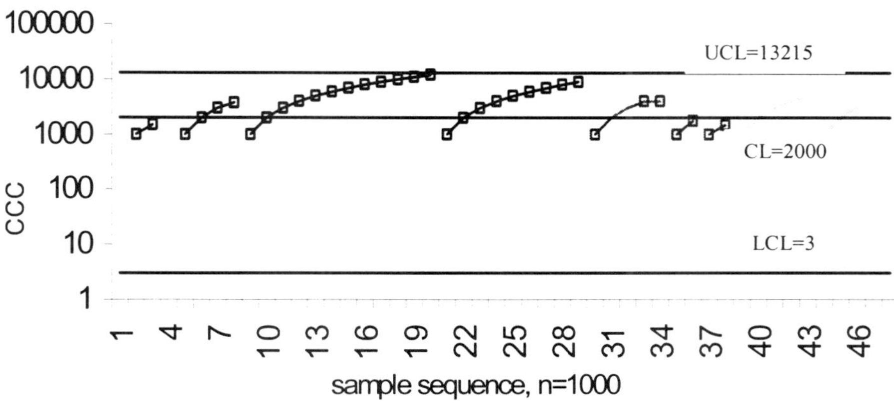

Figure 3.6. A typical layout of a *CCC* chart using Calvin's method.

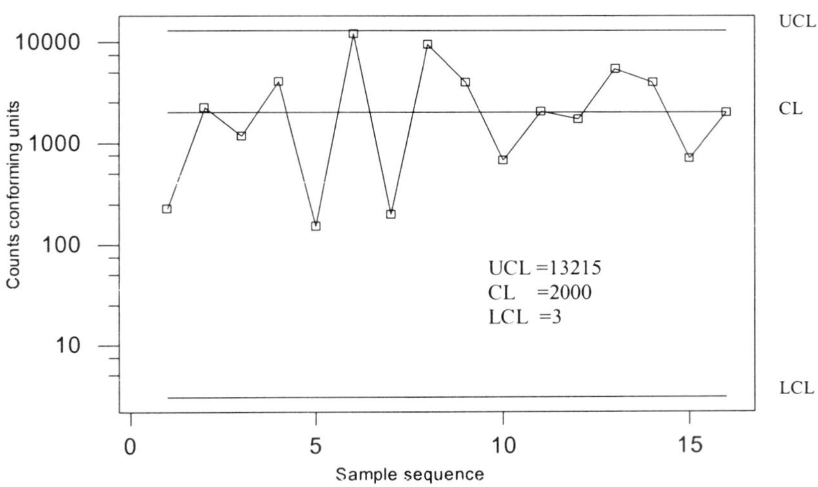

Figure 3.7. The *CCC* chart on semi-log scale which is recommended.

3.5. Some comparisons with 3-sigma limits

Traditional control limits are derived based on the mean and three times the standard deviation. When count data is geometrically distributed, this method could also be used. Kaminsky *et al.* (1992) showed that if the process we being monitoring is best described by geometrically distributed events, wrong decisions are commonly made if the conventional *c*-charts are used. In line with the conventional idea, Kaminsky *et al.* (1992) proposed new control limits for the use of geometric distribution applied to count data and grouped data is used. The control limits are based on three times the standard deviation assuming the underlying distribution to be geometric. Some quantitative comparison with the probability limits is presented here.

If the observation X is geometrically distributed, then we have

$$P(X=x)=p(1-p)^{x-1},\ \ x=1,2,...,i,... \tag{3.16}$$

If we have a subgroup of size n, then the total number of counts in the subgroup, $X=X_1+X_2+...+X_n$, has a negative binomial distribution. Assuming that p is known and n is fixed, control limits can easily be obtained.

Let $Z=X-n$, the negative binomial distribution can be written in this form,

$$P[Z=z]=\binom{n+z-1}{n-1}p^n(1-p)^z,\ \ z=0,1,2,... \tag{3.17}$$

The expected value and the variance of Z are given by

$$E[Z]=n(1-p)/p \tag{3.18}$$

and

$$Var(Z)=n(1-p)/p^2 \tag{3.19}$$

respectively. In general, based on the conventional idea of k-sigma, that is, control limits for Z computed as the k times the standard derivation, the following control limits are obtained:

$$UCL_k = \frac{n(1-p)}{p} + k\sqrt{\frac{n(1-p)}{p^2}} \tag{3.20}$$

and

$$LCL_k = \frac{n(1-p)}{p} - k\sqrt{\frac{n(1-p)}{p^2}} \tag{3.21}$$

Because the k-sigma idea is based on normal approximation, the sample size has to be large in order for the probability of false alarm to be equal to the case in the conventional Shewhart charts. Stated alternatively, it is well-known that for a negative binomial distribution to be approximated by a normal distribution, np has to be large. When p is small, the sample size n has to be very large.

The problem with the LCL is a more serious one. In fact, LCL can almost never be greater than 0. In order for LCL to be positive, we have to have

$$\frac{n(1-p)}{p} > k\sqrt{\frac{n(1-p)}{p^2}} \tag{3.22}$$

The above inequality can be simplified to be

$$n(1-p) > k^2$$

or

$$p < 1 - k^2 / n \tag{3.23}$$

Suppose that k=3 and n=5, which are standard values commonly used, we have p<-0.8, and this can never happen. Also, when we have continuous inspection, n=1, so that even with k=1, we must have p<0 which can never be valid.

A solution is to use the probability limits that have been discussed. The use of exact probability limits is a known concept and has been mentioned in many articles and textbooks on charts for attributes and variables, see e.g., Duncan (1986) and Wetherill and Brown (1991). This is also in line with the statistical process control ideas originally proposed by Shewhart (1931). Their use is more important in the case of geometrically distributed quantities because of the reasons already discussed.

It can be pointed out that by using the exact probability limits, the requirement for the lower control limit to be greater than zero is

$$P[X=0]=p^n<\alpha/2 \tag{3.24}$$

This is easily satisfied in practice. For example, when α is 0.00135 and with a subgroup size of n=5, p has to be greater than 0.267 for LCL to fall below zero. For p<0.267, process changes can be detected even with simple one-point out of control rule.

Now we present some numerical examples to illustrate the use of the control limits and the use of geometric distribution in general. Kaminsky *et al.* (1992) presented a few examples where geometric distribution is more suitable than the Poisson distribution in describing process characteristic. In one example when studying the number of orders on each truck in a central distribution center, p=0.025 and UCL=465.34 while LCL is less than zero. The exact probability limits are 28 and 567, respectively. Hence, if an observation falls below 28, the process most likely has changed.

Example 3.3:

For a data set provided in Kaminsky *et al.* (1992), p=0.2 and the probability limits are 0 and 62. For the sake of simplicity, data in Table 3.5 is grouped and subtracted by 5 because of the shift of one in the original data. The probability upper control limit (Figure 3.8) is compared with the 3-sigma upper control limit of 51. Note again that the 3-sigma lower control limit is less than zero. The upper control limit is 62 compared with that in the original paper of 51. However, for this small data set, all points are within the control limits.

□

Table 3.5. A set of geometric data, from Kaminsky *et al.* (1992)

22	11	14	15	20	12	11	15	11	40	18	27	23	19	8
22	39	9	19	26	27	30	37	16	18	20	12	12	33	25

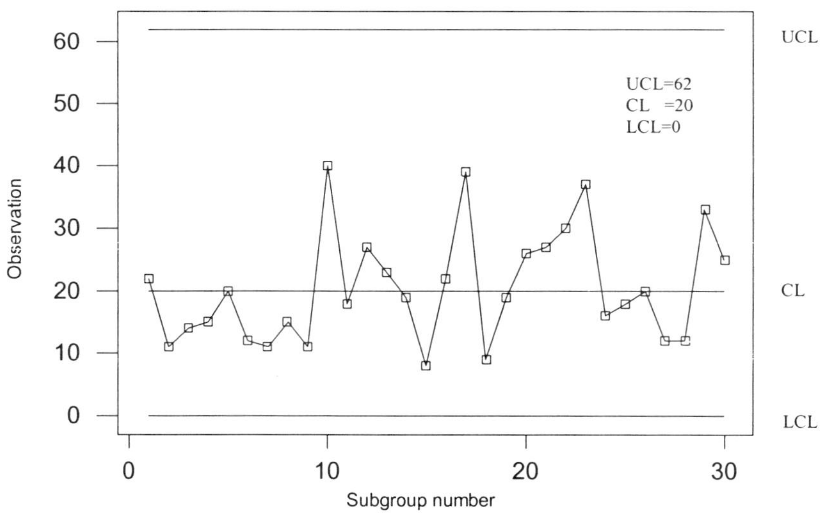

Figure 3.8. A geometric control chart of data set in Table 3.5 with probability limits.

Example 3.4:
To further illustrate the use of probability limits for the *CCC* chart, a set of simulated data is generated based on n=5 and p=0.1. The probability limits are 4 and 139 respectively, while the 3-sigma upper control limit is 109. The data is given in Table 3.6 and the control chart is given in Figure 3.9.

Table 3.6. A set of negative binomial count as the sum of five geometric counts simulated with p=0.1 (read from left-to-right and down)

71	22	88	118	27	37	47	43	39	45
30	105	33	102	49	31	15	38	18	65
61	59	30	73	39	69	34	55	29	69
99	43	38	56	38	28	16	14	106	62
61	24	48	24	48	39	58	20	46	29
46	30	39	62	77	31	43	36	19	22
45	35	20	63	43	37	45	36	68	56
90	14	73	65	50	27	23	60	27	43
36	77	28	81	50	35	67	19	47	41
24	28	28	58	36	61	31	29	62	85

For the data set in Table 3.6, there is one point that falls outside the 3-sigma limit. However, all points are within the probability limits. This case is in line with the result in Section 3.5 indicating that the probability of false alarm is around 1% instead of 0.27% associated with 3-sigma originally.

□

3.6. Cumulative Quantity Control (*CQC*) chart

The *CCC* chart is useful when the process quality is measured in terms of percentage of nonconforming items in the items produced. When the production process is continuous in a sense that production quantity is measured on a continuous scale rather than a count of numbers, a continuous counterpart of *CCC* chart, called *CQC* chart is proposed in Chan *et al.* (2000). This simple extension is briefly described here because of its similarity to the *CCC* chart.

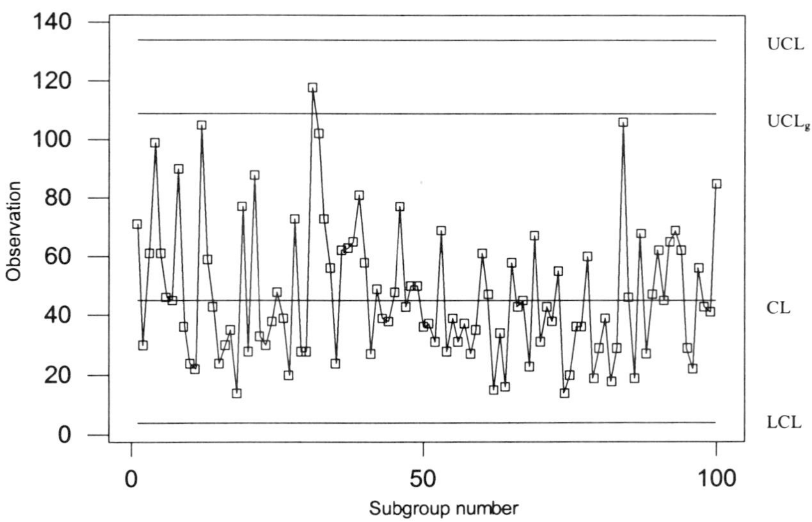

Figure 3.9. The control chart of data in Table 3.6 with UCL_k as the 3-sigma limit =109.

Traditionally, the Shewhart c-chart or u-chart is used to monitor the rate of occurrence of the defects in a process. For processes in which the defect rate is measured in terms of number of defects per unit quantity of product, such as production of steel sheets whose quantity is measured in square meters, liquids in liters, wire in meters, and so on, the traditional c-chart and u-chart cannot be used if the defect rate is very low. The problems similar to those with p- and np-charts described in Section 1.2 occurs. Alternatives are needed for this in a high-quality environment.

For such a case, Nelson (1994) suggested counting the number of defects c observed in a sample, and construct a Shewhart-type control chart for the transformed variable $c^{1/3.6}$. On a control chart for transformed variables, however, direct physical interpretation of the meaning of the original data is not possible.

The cumulative quantity control chart (CQC-chart) operates as follows. The defect occurrence process can usually be modeled by a Poisson process with a mean λ that is the number of defects per unit quantity of product. Then the number of unit(s) Q required to observe exactly one defect is an exponential random variable with parameter λ:

$$F(Q) = 1 - e^{-\lambda Q}, \quad Q \geq 0 \tag{3.25}$$

A control chart for Q can be constructed to monitor the shifts in the process mean. As the cumulative quantity is plotted, the name CQC-chart is used.

In a CQC-chart, a lower control limit (LCL) and an upper control limit (UCL) are defined as the probability limits discussed in chapter 2. They are obtained in such a way that the process is regarded as out of control if the number of unit(s) Q inspected to observe one defect is less than LCL or larger than UCL. Given an acceptable false alarm rate of α, the control limits can be computed as

$$LCL = -\lambda^{-1}\ln(1-\alpha/2) \tag{3.26}$$

and

$$UCL = -\lambda^{-1}\ln(\alpha/2) \tag{3.27}$$

The center line can be defined as the 50-percentile, given by

$$CL = -\lambda^{-1}\ln(1/2) \approx 0.6931\lambda^{-1} \tag{3.28}$$

Such a *CQC*-chart is a double-limit *CQC*-chart, since it has both the lower and upper control limits. In this chart, $Q<LCL$ indicates that the process has probably deteriorated, and $Q>UCL$ indicates that it is likely that the process quality has improved. On a *CQC*-chart, the quantity Q is reset to 0 immediately every time when one defect is observed as for the *CCC* chart.

The implementation, which is similar to the case of a *CCC* chart, has been described in Chan *et al.* (2000). Especially when *CQC* chart is to be used, we need to estimate the value of process levels, λ. This can be estimated as the average of observed Q value from a previous sample.

It should be noted again that the advantages of *CQC* chart, as mentioned before are two fold. It can be used for continuous production process, as well as to monitor both process improvement and deterioration effectively.

CHAPTER 4

Process Improvement Detection

Usually technology advances will lead to process improvement with time and could ultimately approach the states of zero-defect (ZD). Traditional statistical process control tools would find themselves inadequate in providing the needed decision support. New statistical analysis techniques and decision making procedures are therefore continually been evolved, both to handle high performance processes as well as to look for process improvement opportunities, not just to tell possible process deterioration. In this chapter, we discuss some statistical as well as management issues related to process improvement detection. First, the need and potential for continuous improvement even in a near ZD environment is discussed, and then we outline some management and statistical issues. Some potential possibilities are also summarized.

4.1. Potential for process improvement

Continuous process improvement is a catch-phrase in total quality management, and stringent quality requirements in some industries have led to concepts such as Motorola's "six-sigma" measures of performance. Such a transition in levels of performance cannot be realized or sustained without a change in quality management and engineering techniques. Many traditional quality management concepts could be revised.

For example, well published phrases such as "Zero-Defects" and "Do It Right Each time", once regarded as useful ideas and motivators, may not be useful in reality (Goh and Xie 1994). In line with continuous improvement, "Do It Better Each Time" offers a more promising and proactive approach. In fact, different quality philosophies could be integrated for continuous improvement as shown in Figure 4.1.

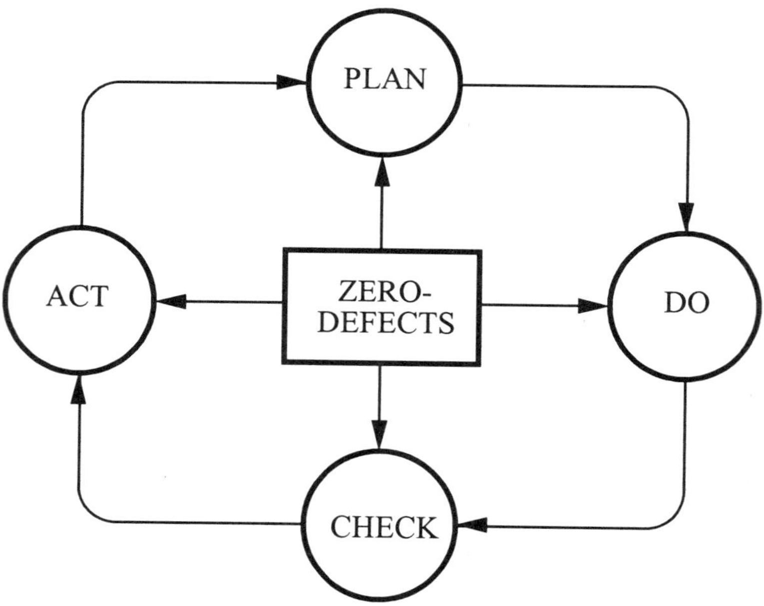

Figure 4.1. The integration of zero-defects concept in Deming's PDCA circle.

4.1.1. Need for improvement detection

Quality improvement efforts can generally be recognized in three categories, namely management, technology and information. Quality management relates to various concepts or steps advocated by quality gurus such as Crosby, Deming, Juran and Taguchi, and systems, such as ISO9000 standards or the recent Six Sigma framework, that must be in place to effect actions at the organizational level.

Quality technology encompasses all hardware and software resources needed for design, development, production, and after-sales activities. Being product-specific, technology selection and technology application are largely an engineering function and will not be addressed here. Quality information techniques enable an organization to evaluate status and measure status changes; statistical tools being at the core of quantitative procedures, they are vital for judicious decision making at the various stages of product realization.

The importance of capabilities in organizational management and application of technologies has been well recognized. The value of measured quality information, on the other hand, has not been necessarily fully exploited in many industrial organizations. In fact, effective decision making via such information is one direction in which quality can be further improved in an organization after its management system and technology tools have been stretched to their maximum effectiveness.

Generally, there is a need for tools to "Do It Better Each Time" and many conventional techniques are not developed for this purpose. With the shift in the emphasis and the need for further quality improvement once a ppm (parts per million) or ppb (parts per billion) level of performance has been attained, statistical tools have to be developed or modified and this is the main theme of this Chapter.

4.1.2. Problems associated with improvement detection in near ZD environment

In a near ZD environment, process improvement cannot be easily detected with the traditional attribute charts. Some statistical problems are mentioned in Section 1.2 as the lower control limits tend to be less then zero. In fact, even when probability limits are used, this problem remains. A better approach in a near ZD environment is to use the *CCC* chart introduced in Chapter 3 and more discussions on this will be discussed in Section 4.2. Here we mainly focus on some practical and managerial issues.

Indeed, there are many problems associated with process improvement. Without management involvement and encouragement, this cannot be done in a near ZD manufacturing environment. Some statistical issues have been pointed out before and will not be repeated here. We discuss below some management issues related to improvement detection.

(a) Problems with defining improvement

Although it is straightforward to define what an improvement is by using quantitative measurements, it is not always the case that this can be done without problems. If the process characteristics are not related to what the customer mean by process quality, then this can only be a wasted effort.

(b) Problems with measuring improvement

Even when the process characteristics in line with customer needs are well-defined and measurable, it is usually a problem in a near ZD environment that we are not sure with what degree of confidence can we judge the process having improved. Control charts have to be correctly used and decisions have to be made with a statistical confidence afforded by the data.

(c) Problems with improvement actions

When an improvement potential is identified, a decision has to be made to improve the process. This is not straightforward in practice and some cost factors have to be assessed. If an improvement is due to the use of more expensive process elements such as material or personnel, then management has to evaluate the cost-effective of possible options.

(d) Problems with sustaining improvement

It is usually suggested that when a process is improved, control limits have to be revised to reflect the new state of the process. However, it takes time to know whether the process improvement has been sustained and if not, the early charts still have to be used. Sustaining improvement has to be a consideration in any improvement effort. Having done it "right the first time" does not mean right the next time, and definitely not "right forever".

(e) Problems with further improvement

According to Deming's PDCA cycle, whenever a process has reached an improved state, the next step is to plan for further improvement. This is a never-ending effort and any signal for improvement should not be missed. However, this is usually difficult and much more process-related information is needed. The use of *CCC* chart presented in Chapter 3, on the other hand, could provide much needed information for detecting the potential for further improvement.

4.1.3. Some possibilities for improvement detection

When properly used, control charts are useful for detecting process improvement as well as detecting process deterioration. It seems that there has been a missing link between the use of SPC and continuous improvement. One of the goals of this book to bridging this gap.

Our general strategy is not to base any judgment on findings from individual samples, but to make use of a continuous stretch of data from the immediate past, i.e. track record of the process in question. This in turn raises a very pertinent point: if track record and current performance data could alert their user significant slips of process performance, why not make use of them to look for process improvement?

Just as an on-going process can be adversely affected by fortuitous circumstance, occasionally an exceptionally superior performance could also materialize which, if unnoticed, would mean the loss of a valuable opportunity to identify and sustain a special set of conditions, i.e. "assignable causes" in the parlance of statistical process control, that can lift long-term quality performance to a higher plateau.

Some technical details, evolved from process control charting procedures, have been formulated for process improvement detection in this book. The focus is actually more on the necessity for new decision making principles in the age of near ZD performance than for their mathematical novelty. In short, statistical techniques can and should be used for the quest for improvement, rather than just playing the traditional role of detection of deterioration in performance.

Although many buzzwords should not be used on the surface, most quality philosophies can play a constructive role. Each has its own limitations. To make a better use of them to motivate continuous improvement and to put them in action, an integration of various total quality management philosophies is needed. For example, a linkage between PDCA and ZD (Figure 4.1) can be established based on the Do It Better Each Time (DIBET) philosophy. Each PDCA cycle consists of a number of activities. The ZD ideal can be present in these activities, motivating people to do their best. In practice, the application of DIBET involves the following steps:

(1) Set a reasonable goal;

(2) Try to reach the goal;

(3) Take quantitative measurements and conduct quantitative analysis;

(4) Carry out the improvement.

These steps are in line with the PDCA cycle and should be repeated over and over again. In each phase of a PDCA cycle, there is a need to have the ZD philosophy in mind and preventive measures should be taken with ZD as the target.

4.2. Using *CCC* chart for process improvement

The *CCC* chart discussed in Chapter 3 is a powerful charting technique when the process is near ZD and samples are mostly conforming. It can be noted that the *CCC* chart is also very useful for detecting process improvement, which the conventional charting techniques would fail even with the common run rules. Although the use of *CCC* chart has been described in Chapter 3, in this section we look at the details of decision making and the revision of control limits.

Suppose a *CCC* chart is constructed following the procedures in Chapter 3. Control limits are derived as before. As indicated previously, whenever a point falls below the *LCL*, the process has probably deteriorated, as there are too few conforming items between two nonconforming ones. However, if a point falls above the *UCL*, it is likely that there are causes that have made the process improved. In this case, the process can be studied and whenever a reason for the improvement is found, effort should be made to sustain it.

Recall that for the *CCC* chart, the *UCL* is given by

$$UCL = \frac{\ln(\alpha/2)}{\ln(1-p)} \tag{4.1}$$

and for any level of fraction nonconforming the *UCL* can easily be determined given an acceptable false alarm probability. Whenever the cumulative number of nonconforming items has exceeded this value, positive assignable causes should be searched; whenever possible, such causes should be maintained rather than removed since the process will be running at a higher level of quality.

For the improved process, the control limits have to be revised. However, it is advisable to keep the control limits until about 20 points are obtained, before they are revised upwards. A plot similar to that in Figure 4.2 illustrates such a situation.

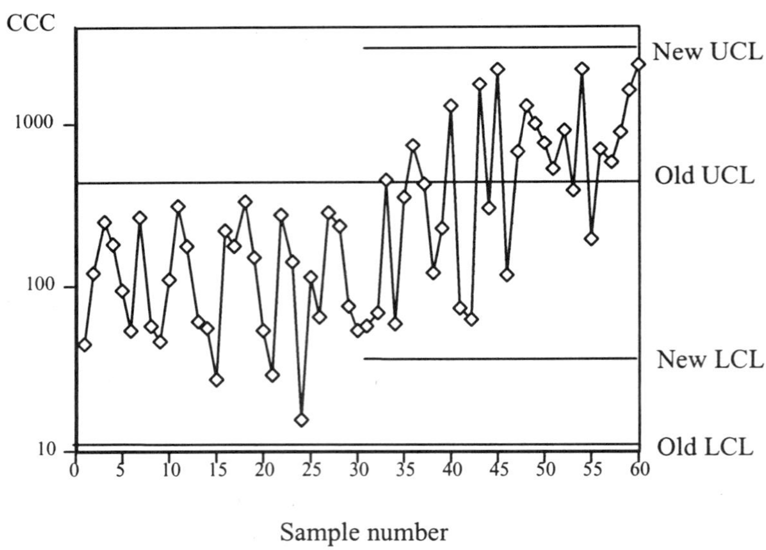

Figure 4.2. *CCC* chart and the revision of the limits when the process is improved and the improvement is sustained.

From this, we can see that the use of *CCC* chart makes it possible to detect process improvement even in a ZD environment. In fact, the resolution of the *CCC* chart on a log-scale is higher when the process quality is higher.

4.2.1. Use of *CQC* chart

Similar to the use of *CCC* chart, the *CQC* chart discussed in Section 3.5 can also be useful in detecting process improvements when the process is continuous in nature. Recall that the *CQC* chart monitors the quantity till the observation of a defect or a nonconforming item. Note that the quantity here does not have to be an integer. The *UCL* of *CQC* chart is given by

$$UCL = -\lambda^{-1} \ln(\alpha / 2) \tag{4.2}$$

and when a point falls above *UCL*, it indicates that the process has improved. This is because it has taken a much longer time or an excessively large quantity before a defect is observed. In this case, assignable cause should be found, and whenever possible, the cause should be maintained so that the process remains at the high-quality level.

It should be noted that when implementing *CCC* or *CQC* for improvement detection, it is not necessary to wait until a point is plotted on the chart. As long as the cumulative count or cumulative quantity has exceeded the respective *UCL*, a positive signal is evident and investigation could be started.

CHAPTER 5

Modified Implementation of Geometric Chart

In this chapter some important issues on the geometric chart are discussed. Although the geometric charts are shown to be suitable for the monitoring of high quality processes, it also has some problems since the standard assumptions are similar to those of the traditional Shewhart chart. Some modifications are necessary for more accurate interpretations. Although it is not possible to eliminate all potential problems, we could minimize the risk of encountering problems by adjusting the operating procedure or optimizing the control limits, as suggested in the following sections.

We first discuss a problem with the average run length (*ARL*), the expected number of points plotted before an alarm is raised. Due to the skewness of geometric distribution, the *ARL* curve does not have its maximum at the process level used to compute the control limits. This implies that when process is shifted slightly, it will take a longer time to raise an alarm. A procedure is developed in Section 5.1 to deal with this problem. Furthermore, some transformations are discussed in Section 5.2. A conditional procedure making use of the information of earlier counts is presented in Section 5.3. Finally the problems of inspection and estimation errors, which have significant impact on any control charting procedures, are discussed in Section 5.4 and 5.5, respectively.

5.1. Optimum limits for maximum average run length

The average run length, which is here defined as the expected number of points plotted on a control chart before an out-of-control signal is observed, is an important characteristic of a control chart. The *ARL* should be large when the process is considered to be in control or when operating at the prescribed level. When the level has shifted, the average should decrease so that alarms can be raised quickly.

5.1.1. *ARL* of the basic *CCC* chart

For the geometric chart with given fraction nonconforming p, the *ARL* in terms of the average number of points plotted before an alarm is raised, is

$$ARL = \frac{1}{1-\beta} = \frac{1}{1-(1-p)^{LCL} + (1-p)^{UCL}} \tag{5.1}$$

where β is the Type II error or the probability of no alarm when the process has shifted. When p with a value p_0 is the level of acceptable quality, or alternatively, the average quality level is estimated from the process, the *ARL* is $1/\alpha$ where α is the false alarm probability. Note that for *CCC* chart, this is the same as the number of nonconforming items observes when an alarm is raised.

Although control charts based geometric distribution has been shown to be useful in high yield manufacturing process monitoring and other applications, the average time to alarm may initially increase when the process has deteriorated. That the maximum of the *ARL* is not found at the nominal process level is a common phenomenon for skewed distribution (Xie *et al.*, 2000a).

It is pointed out in Xie and Goh (1997) that the traditional 3-sigma limits such as those in Kaminsky *et al.* (1992) should not be used as the geometric distribution is always skewed and the normal approximation is not valid. In fact, in most of the literature on *CCC* chart, exact probability limits have been adopted. However, although this will provide a fixed false alarm probability, the average time to an alarm when the process is deteriorated, in terms of average run length computed by Eq.(5.1), may increase in the beginning.

The *ARL* is highly sensitive in identifying shifts quickly when a process has deteriorated or improved. If the process is stable or in control, we will experience a false alarm signal in about $1/\alpha$ of the samples or points plotted on the control chart. For α=0.0027, the *ARL* is 370 which should be constant irrespective of process average. However, for geometric chart, the *ARL* may initially increase when the process has deteriorated. For instance, for a process average of 50ppm, the *ARL* is 370 but the maximum *ARL* occurs only at p is 505. Suppose the process is shifted from 50ppm to 60ppm; then the *ARL* is 505 which is much higher than the normal of 163, which is the ARL of 40ppm.

In Table 5.1, some *ARL* values for p=50ppm and α=0.0027, 0.005 and 0.010 are shown. The *ARLs* are significantly lower at p=50ppm irrespective of false alarm probabilities and they are attained maximum after 50ppm. In fact the maximum occurs at about one and quarter times of the desired p value, which can be denoted as p_1. Figure 5.1 shows this undesirable area under the *ARL* curve between the points $ARL(p_0)$ and $ARL(p_1)$.

Assume that p=50ppm is shifted to 70ppm, or a 40 percent deterioration, then at least 504 points or samples are required to detect the shift. That is, if one item is inspected every minute, it will take 504 minutes for a geometric chart to raise an out of control signal. Before the alarm, the process, though deteriorated, is still considered to be in control.

This implies in practice that it could take a longer time to alarm when the process is deteriorated than when the process is normal even in the case of using the exact probability limits. To eliminate this problem, a new procedure using an optimization procedure is developed in Xie *et al.* (2000a) to obtain the maximum *ARL* at the desired value of *p*. The procedure is described in the following.

Table 5.1. Some *ARL* values at process average 50ppm for various α

Fraction Non-Conforming	α=.0027	α=.005	α=0.01
1	1	1	1
10	4	3	3
20	14	11	8
30	51	35	22
40	163	97	54
50	370	200	100
60	505	266	129
70	504	268	132
80	458	246	122
90	411	221	110
100	370	200	100
110	337	182	91
120	309	167	84
140	265	143	72
160	232	125	63
180	206	111	56
200	186	100	50

5.1.2. Procedure to maximize *ARL* at a desired process level

By adjusting the control limits properly, the undesirable area could be eliminated or minimized. An optimization method either to minimize this undesirable area or to maximize the *ARL* at the desired process average is proposed in Xie *et al.* (2000a).

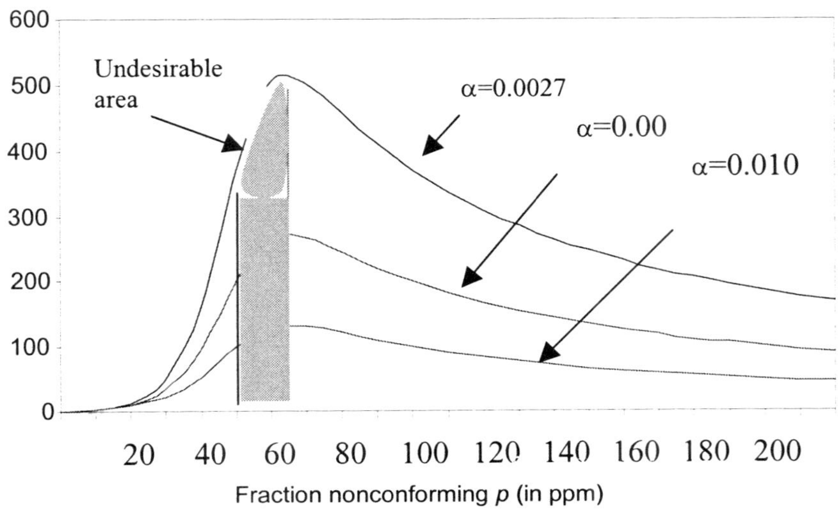

Figure 5.1. *ARL* curves with p=50ppm for α=.0027, .005 and .010 with the maximum *ARL* at a higher p value.

Using Eq.(5.1), the maximum *ARL* at $p=p_0$ can be obtained by maximizing the *ARL* function. By differentiating the function and then solve for p, we get

$$(1-p_0)^{L-1}\ln(1-p_0)\frac{dL}{dp}-(1-p_0)^{U-1}\ln(1-p_0)\frac{dU}{dp}=0 \qquad (5.2)$$

where L is the LCL and U is the value of UCL.

After simplifying the above equation, we get the p value where *ARL* reaches the maximum:

$$p = 1 - \exp\left\{ \frac{\ln(1-p_0)\ln\left[\frac{(\alpha/2)}{(1-\alpha/2)}\right]}{\ln\left[\frac{\ln(1-\alpha/2)}{\ln(\alpha/2)}\right]} \right\} \tag{5.3}$$

Substituting this function into Eq.(5.3), the new control limits are obtained:

$$LCL = \gamma_\alpha \frac{\ln(1-\frac{\alpha}{2})}{\ln(1-p_0)} \tag{5.4}$$

and

$$UCL = \gamma_\alpha \frac{\ln(\alpha/2)}{\ln(1-p_0)} \tag{5.5}$$

where

$$\gamma_\alpha = \frac{\ln\left[\frac{\ln(1-\alpha/2)}{\ln(\alpha/2)}\right]}{\ln\left[\frac{(\alpha/2)}{(1-\alpha/2)}\right]} \tag{5.6}$$

The equations above do not have a direct probability interpretation. However, the new control limits can be derived by multiplying the probability limits with a constant γ_α (adjustment factor). Since the adjustment factor γ_α is purely a function of false alarm probability, the original process average remains unchanged for the new limits.

In particular, because of this simple adjustment, the original scale is never affected except that the existing control limits are shifted by 'γ_α' times. A set of values is also presented in Table 5.2. A similar procedure has been proposed in Chen (1998) for the *p*-chart by simple adjustments on the limits and for which the Cornish-Fisher expansion of quantiles was used to obtain a better normal approximation.

Table 5.2. Some values of adjustment factor, γ_α computed with Eq.(5.6) for different false alarm probability, α

α	γ_α	α	γ_α	α	γ_α	α	γ_α
0.0001	1.2315	0.0026	1.2852	0.0051	1.2996	0.0076	1.3088
0.0002	1.2411	0.0027	1.2859	0.0052	1.3000	0.0077	1.3091
0.0003	1.2471	0.0028	1.2867	0.0053	1.3005	0.0078	1.3094
0.0004	1.2515	0.0029	1.2874	0.0054	1.3009	0.0079	1.3097
0.0005	1.2551	0.0030	1.2881	0.0055	1.3013	0.0080	1.3100
0.0006	1.2581	0.0031	1.2888	0.0056	1.3017	0.0081	1.3103
0.0007	1.2607	0.0032	1.2895	0.0057	1.3021	0.0082	1.3106
0.0008	1.2630	0.0033	1.2901	0.0058	1.3025	0.0083	1.3109
0.0009	1.2650	0.0034	1.2907	0.0059	1.3029	0.0084	1.3112
0.0010	1.2669	0.0035	1.2914	0.0060	1.3033	0.0085	1.3115
0.0011	1.2686	0.0036	1.2920	0.0061	1.3037	0.0086	1.3118
0.0012	1.2702	0.0037	1.2925	0.0062	1.3040	0.0087	1.3121
0.0013	1.2717	0.0038	1.2931	0.0063	1.3044	0.0088	1.3123
0.0014	1.2730	0.0039	1.2937	0.0064	1.3048	0.0089	1.3126
0.0015	1.2743	0.0040	1.2942	0.0065	1.3051	0.0090	1.3129
0.0016	1.2756	0.0041	1.2948	0.0066	1.3055	0.0091	1.3132
0.0017	1.2767	0.0042	1.2953	0.0067	1.3058	0.0092	1.3134
0.0018	1.2778	0.0043	1.2958	0.0068	1.3062	0.0093	1.3137
0.0019	1.2789	0.0044	1.2963	0.0069	1.3065	0.0094	1.3140
0.0020	1.2799	0.0045	1.2968	0.0070	1.3069	0.0095	1.3142
0.0021	1.2809	0.0046	1.2973	0.0071	1.3072	0.0096	1.3145
0.0022	1.2818	0.0047	1.2978	0.0072	1.3075	0.0097	1.3147
0.0023	1.2827	0.0048	1.2982	0.0073	1.3079	0.0098	1.3150
0.0024	1.2835	0.0049	1.2987	0.0074	1.3082	0.0099	1.3152
0.0025	1.2844	0.0050	1.2991	0.0075	1.3085	0.0100	1.3155

Example 5.1:

In the example, the data set of Xie *et al.* (1995) is used and also presented as Table 5.3 for p=500ppm and a false alarm probability α=0.10. For this example, the original control limits are 102 and 5989 and the proposed limits are 141 and 8274. Using these limits, the simulated values are plotted in Figure 5.2.

From the chart, it is observed that two false alarm signals are encountered with the proposed limits whereas the original method indicated three. Similarly, in identifying process shift, both methods show the same *ARL* when the process is shifted to 50ppm. This indicates that the proposed method has a smaller number of false alarms over the existing method. The proposed procedure is easily implemented and no complex calculation is needed compared with existing method except the use of an adjustment factor.

□

5.2. Charts based on transformations

In this section, transformation based on geometric distribution is considered since symmetric limits are of interest in practice. The transformation is useful when the distribution is non-normal which is the case for geometric distribution. Under the transformation, the data become normally distributed. There are a number of transformations available in the literature, and in this section a few of them are shown.

5.2.1. Some transformation approaches

Quesenberry's *Q*-Transformation

Using ϕ^{-1} to denote the inverse function of the standard normal distribution, define

$$Q_i = -\phi^{-1}(u_i) \tag{5.7}$$

where $u_i = F(x_i; p) = 1-(1-p)^{x_i}$. For i=1,2,.., Q_i will approximately follow the standard normal distribution, and the accuracy improves as p approaches zero.

Table 5.3. A set of geometric quantities for given p=500ppm and α=0.1

Nonconforming Items	*ARL*	Simulation Legend
1	3706	p =500ppm
2	9179	p =500ppm
3	78	p =500ppm
4	1442	p =500ppm
5	409	p =500ppm
6	3812	p =500ppm
7	7302	p =500ppm
8	726	p =500ppm
9	2971	p =500ppm
10	42	Shift, p =5000ppm
11	3134	p =5000ppm
12	1583	p =5000ppm
13	3917	p =5000ppm
14	3496	p =500ppm
15	2424	p =500ppm
16	753	p =5000ppm
17	3345	p =5000ppm
18	217	p =5000ppm
19	3008	p =500ppm
20	3270	p =500ppm
21	5074	Shift, p =50ppm
22	3910	p =50ppm
23	23310	p =50ppm
24	11690	p =50ppm
25	19807	p =50ppm
26	14703	p =50ppm
27	4084	p =50ppm
28	826	p =50ppm
29	9484	p =50ppm
30	66782	p =50ppm

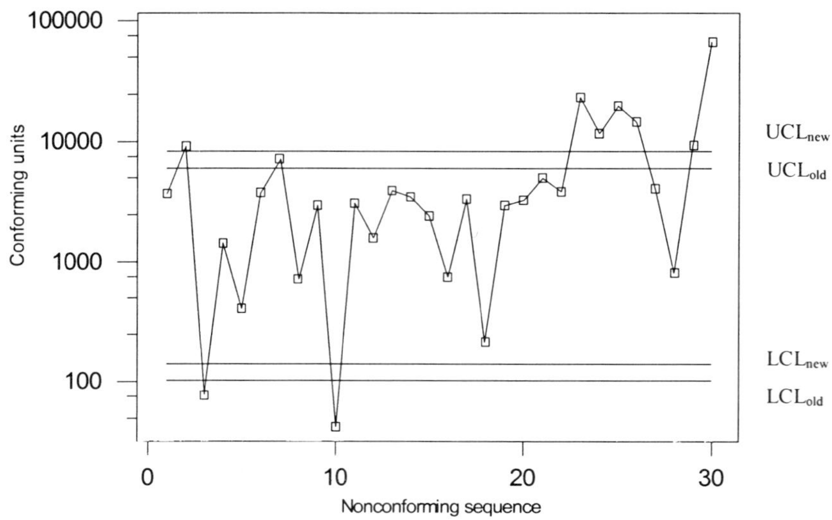

Figure 5.2. The *CCC* chart showing both the probability limits and the modified limits.

Theoretically, the *Q*-transformation should serve the purpose well as it is based on the exact inverse normal transformation. A practical problem is that the model parameter has to be assumed known or accurately estimated and the result could be sensitive to erroneous estimates. A sensitivity study of this issue will be carried out in a later section to further explore this issue. Also, it is not practical to use the inverse normal transformation if it is not already built in.

The log transformation

A geometric distribution is a special case of negative binomial distribution. For a negative binomial variable z with mean m and exponent k, we have that

$$p(z=r)=\frac{(r+k-1)!}{r!(k-1)!}\left(\frac{m}{m+k}\right)^r\left(\frac{k}{m+k}\right)^k; \quad r=0,1,2,\ldots,\infty \tag{5.8}$$

Anscombe (1948) showed that the corresponding transformation,

$$y=\sinh^{-1}\sqrt{\frac{z+c}{k-2c}} \tag{5.9}$$

can be used to transform a negative binomial distribution into a normal one. Anscombe (1948) also showed that a similar transformation, known to have an optimum property for large m and $k \geq 1$, is

$$y=\ln(z+k/2) \tag{5.10}$$

It can be shown (Xie *et al.,* 2000b) that when z is large, which is for example the case when a high quality process is monitored based on cumulative counts of conforming items between two non-conforming ones, both the inverse hyperbolic and the log transformation are equivalent to the simple logarithmic transformation.

A double square root transformation

A simple transformation is the double square root (or fourth root) transformation

$$y=x^{1/4}; \qquad x\geq 0 \tag{5.11}$$

It is well known that the square-root transformation can be used to convert positively skewed quantities to normal. However, the geometric distribution is so skewed, so that even after the square root transformation it is still positively skewed. Although based on the idea of transforming exponential to normal, a power transformation of $x^{1/r}$ for a r>3.6 as given by Nelson (1994) should be used, considering the accuracy and simplicity (Xie *et al.* 2000b), the double square root transformation is preferred as it can be easily performed.

Our simulation studies have shown that the results of double square root transformation generally perform very well.

Box-Cox transformation

Box and Cox (1964) presented a procedure that can be applied to transform non-normal distribution to a normal one. For a nonnegative random variable, the Box-Cox power transformation is defined as

$$y(\lambda) = \begin{cases} (x^{\lambda} - 1)/\lambda, & \lambda \neq 0 \\ \log x, & \lambda = 0 \end{cases} \tag{5.12}$$

The estimation of λ is usually a problem and it requires considerable computational effort. On the other hand, since the geometric distribution has exponential distribution as the continuous counterpart, and it is shown in Yang and Xie (2000) that the exponential variable can be transformed to near normality by a power of 0.2654, this is very similar to the double square root transformation which is very simple.

In general, the Q-transformation is based on the exact probability distribution and it should provide the best transformation to normal distribution. However, the parameters, and hence the distribution, is usually unknown. The double square root transformation, despite of its simple form, has shown to be good for geometric distribution. An example is shown below.

5.2.2. An application example

In this section, an example is shown to illustrate some possible transformation approaches. Some further analysis, such as sensitivity study is also carried out to illustrate the use of the method.

Example 5.2:
A data set from Quesenberry (1995) shown in Table 5.4 is used to compare among the all transformations. The corresponding charts for double square transformation and Q-transformation are show in Figure 5.3 and Figure 5.4, respectively.

Table 5.4. A set of geometrically distributed data from Quesenberry (1995) with values of transformations

Obs.	$X^{1/4}$	ln X	Q	Obs.	$X^{1/4}$	ln X	Q
18	2.0598	2.8904	-0.2605	13	1.8988	2.5649	0.0334
1	1.0000	0.0000	1.6448	4	1.4142	1.3863	0.8946
19	2.0878	2.9444	-0.3124	64	2.8284	4.1589	-1.7801
32	2.3784	3.4657	-0.8643	11	1.8212	2.3979	0.1733
28	2.3003	3.3322	-0.7133	12	1.8612	2.4849	0.1013
8	1.6818	2.0794	0.4218	2	1.1892	0.6931	1.2959
4	1.4142	1.3863	0.8946	21	2.1407	3.0445	-0.4109
27	2.2795	3.2958	-0.6734	34	2.4147	3.5264	-0.9352
72	2.9130	4.2767	-1.9617	15	1.9680	2.7081	-0.0921
8	1.6818	2.0794	0.4218	19	2.0878	2.9444	-0.3124
41	2.5304	3.7136	-0.1646	10	1.7783	2.3026	0.25
75	2.9428	4.3175	-2.0267	6	1.5651	1.7918	0.6282
29	2.3206	3.3673	-0.7523	7	1.6266	1.9459	0.5196
4	1.4142	1.3863	0.8946	107	3.2162	4.6728	-2.6408
11	1.8212	2.3979	0.1733	29	2.3206	3.3673	-0.7523
30	2.3403	3.4012	-0.7904	4	1.4142	1.3863	0.8946
32	2.3784	3.4657	-0.8643	36	2.4495	3.5835	-1.0036
15	1.9680	2.7081	-0.0921	1	1.0000	0.0000	1.6448
29	2.3206	3.3673	-0.7523	20	2.1147	2.9957	-0.3625
10	1.7783	2.3026	0.25	23	2.1899	3.1355	-0.5033
25	2.2361	3.2189	-0.5906	28	2.3003	3.3322	-0.7133
4	1.4142	1.3863	0.8946	24	2.2134	3.1781	-0.5475
10	1.7783	2.3026	0.25	55	2.7233	4.0073	-1.5586
20	2.1147	2.9957	-0.3625	7	1.6266	1.9459	0.5196
19	2.0878	2.9444	-0.3124	3	1.3161	1.0986	1.0686

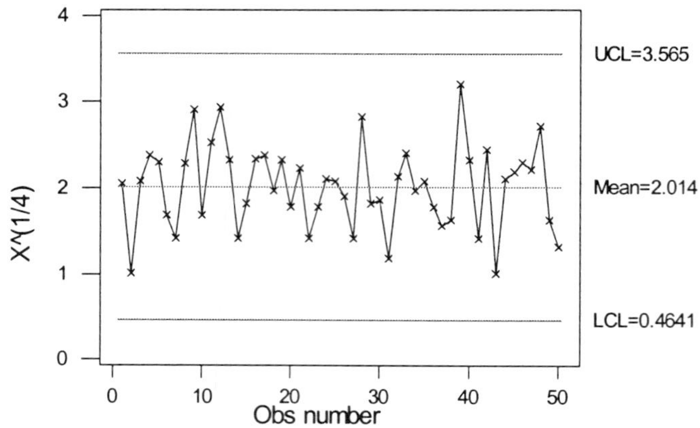

Figure 5.3. The individual chart of *CCC* data with double square root transformation.

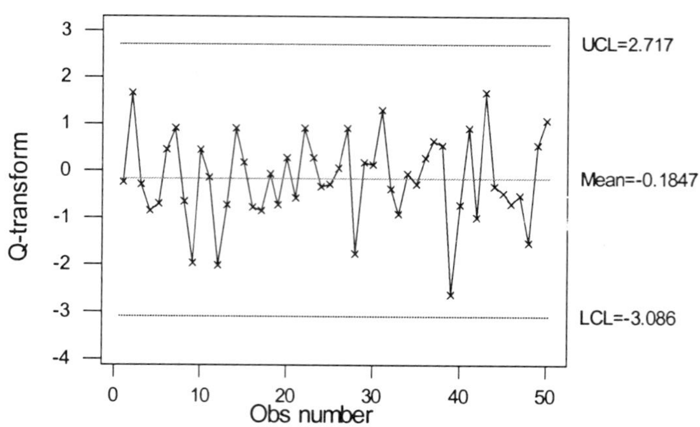

Figure 5.4. The individual chart of *CCC* data with *Q*-transformation.

The Anderson-Darling (*AD*) test statistic and some other non-parametric statistics are used to test the normality of the distribution after the transformation. The results are presented in Xie *et al.* (2000b) and are shown here in Table 5.5.

Table 5.5. Comparison of three transformations using the data in Table 5.4

	Skewness	Kurtosis	*AD* Statistic
$X^{1/4}$	0.1	2.8	0.30
$\ln X$	-0.6	3.1	0.69
Q-transform	-0.2	3.0	0.31

It can be seen from the Anderson-Darling test that the normality hypothesis is not rejected for the *Q*-transformation and double square root transformation but is rejected for the log transformation. Furthermore, the scores for the former two are very close.

□

A sensitivity study

The *Q*-transformation should be the most appropriate transformation as it is based on the exact probability values. However, the parameter of the distribution is assumed known and it would be interesting to see how the *Q*-transformation is affected by errors in the value of p. Recall that the *Q*-statistic is defined as

$$Q_i = -\phi^{-1}(u_i), \quad i = 1,2,... \tag{5.13}$$

where

$$u_i = F(x_i; p) = 1 - (1-p)^{x_i} \tag{5.14}$$

When $p=p_0$ is constant, the Q-transformation can give independent and approximately standard normal statistics. It can be noted that because of the discrete nature the approximate normality depends on p_0. Furthermore, a smaller p_0 will give better approximations as it is closely approximated by exponential distribution, which is continuous. A sensitivity study is illustrated with an example in the following.

Assume that p_0=0.01 for a process. Using different p-values, the percentage change in Q-value can be calculated as

$$\frac{\Delta Q}{Q}=\frac{Q_0-Q_1}{|Q_0|} \tag{5.15}$$

where Q_0 represents the Q-statistic using p_0 and Q_1 represents the Q-statistic using p_1. Some numerical values are shown in Table 5.6.

Table 5.6. Sensitivity analysis of Q when p is changed

$\Delta p/p$	$\Delta Q/Q$	$\Delta p/p$	$\Delta Q/Q$
10%	26%	-10%	-24%
20%	32%	-20%	-49%
30%	115%	-30%	-62%
40%	199%	-40%	-88%
50%	217%	-50%	-122%
60%	296%	-60%	-153%

It can be noted that the value of Q-transformation can be very sensitive to the change in p-value. In practice, the exact value of p is not known. This means that a change in p by a factor of 50 percent is not rare, especially for the geometric chart when it is used in high yield process control.

Because of Q-transformation is relatively sensitive to the error in p-value, the same p-value should not be used when the actual p has changed or when the p-values are not the same for different processes. However, it is not easy to use different p-values to transform the data in different cases. This is another limitation of the Q-transformation and it makes the simple transformation more practical. The double square root transformation and even log-transformation might be easy to use from a practical point of view.

5.3. A conditional decision procedure

As for the traditional implementation of Shewhart chart, a decision when using the original *CCC* chart is based on a single count value. This could be relatively insensitive to small process shift. In this section, a conditional procedure (Kuralmani *et al.* 2002) is described whereby the sensitivity is improved when the process is shifted from the intended level with the false alarm signal fixed at the acceptable process level. The idea of a conditional procedure is to utilize some of the previous runs when a count value exceeds the limits. This procedure is also similar to the supplementary run rules (Nelson, 1984 and Champ and Woodall, 1987) associated with Shewhart charts, but the conditional chart reduces the Type I error and the supplementary run rules reduce the Type II error.

The conditional procedure operates as follows. We count the conforming units until a nonconforming unit is found. Then it is concluded that the process is in control if

1. the count of conforming units are within lower and upper control limits or
2. k previous runs were in control even if the count of conforming units are not within the control limits

5.3.1. Determination of control limits for fixed k

To implement the conditional procedure, k is usually a fixed constant. The choice of k, which should be small, will be discussed later. For a fixed k, the control limits can be derived as in the following (Kuralmani *et al.*, 2002).

For a fixed value of p and a prescribed false alarm probability α, the conditional limits are given as:

$$LCL_{cond} = \frac{\ln[1-(1-\delta)/2]}{\ln[1-p]} \tag{5.16}$$

and

$$UCL_{cond} = \frac{\ln[(1-\delta)/2]}{\ln[1-p]} \tag{5.17}$$

where δ is the solution of the following equation:

$$1-\alpha = \delta + (1-\delta)\delta^k \tag{5.18}$$

The value of δ is in fact the probability that a point is inside control limits in State 1, which is P_{inc} for fixed control limits. It only depends on the preset false alarm probability. It can be seen that δ is smaller than 1-α, which means that the count will be out of the control limits more often. However, as not all points outside of the limits are considered to be out of control point, the false alarm probability can be maintained. Some numerical values of δ are shown in Table 5.7. When k approaches infinitely, it can be seen that δ will approach 1-α. This means that the *CCC* chart is an asymptotic case of the conditional procedure.

Using the values of δ, the control limits for the conditional procedure can be calculated for any given process average with a predetermined false alarm probability. Table 5.8 presents some numerical values.

Table 5.7. Values of δ for various values of α and k

Conditional Number, k	False alarm probability, α		
	0.0027	.01	.05
1	0.94800	0.90000	0.77640
2	0.96290	0.92800	0.83500
3	0.96950	0.94050	0.86150
4	0.97350	0.94800	0.87740
5	0.97620	0.95310	0.88820
6	0.97820	0.95690	0.89620
7	0.97975	0.95990	0.90245
8	0.98100	0.96228	0.90744
9	0.98205	0.96422	0.91155
10	0.98293	0.96590	0.91504
15	0.98592	0.97151	0.92660
20	0.98770	0.97490	0.93324
25	0.98890	0.97720	0.93755
50	0.99191	0.98276	0.94657
50	0.99191	0.98276	0.94657
75	0.99803	0.98517	0.94900
100	0.99830	0.98653	0.94971
200	0.99557	0.98882	0.95000

5.3.2. A study of chart performance

Here we summarize the statistical performance for the conditional procedure. The *ARL* is normally the reciprocal of the in-control probability when the events are independent. In this case the events are not independent, and the unconditional event may be defined as

$$B_i = \{Y_i > UCL \text{ or } Y_i < LCL\} \tag{5.19}$$

where Y_i is the ith observation.

Table 5.8. Values of *LCL/UCL* for conditional charts with α=0.0027

k	Limits	Process average p					
		0.00001	0.00002	0.00005	0.00010	0.00020	0.00050
1	*LCL*	2634	1317	527	263	132	53
	UCL	364964	182481	72991	36495	18246	7297
2	*LCL*	1872	936	374	187	94	37
	UCL	398727	199362	79744	39871	19934	7973
3	*LCL*	1537	768	307	154	77	31
	UCL	418315	209157	83661	41830	20914	8364
4	*LCL*	1334	667	267	133	67	27
	UCL	432374	216186	86473	43235	21617	8645
5	*LCL*	1197	599	239	120	60	24
	UCL	443119	221559	88622	44310	22154	8860
6	*LCL*	1096	548	219	110	55	22
	UCL	451897	225947	90378	45188	22593	9036
8	*LCL*	955	477	191	95	48	19
	UCL	465644	232821	93127	46562	23280	9311
10	*LCL*	857	429	171	86	43	17
	UCL	476356	238177	95269	47633	23816	9525
15	*LCL*	706	353	141	71	35	14
	UCL	495612	247805	99120	49559	24778	9910
20	*LCL*	617	308	123	62	31	12
	UCL	509128	254563	101824	50910	25454	10180

Since Y_i depends on the previous Y_i's, the event B_i is not necessarily independent. However, the event B_i can be defined as a conditional event for given k such that

$$B_i = \{Y_i > UCL(k) \text{ or } Y_i < LCL(k)\} \text{ for all } i \tag{5.20}$$

Then $P(B_i)$ is the actual alarm rate (*AR*), which becomes the false alarm rate (*FAR*) when $p=p_0$. The alarm rate is

$$P(B_i) = P\{Y_i > UCL(k) \text{ or } Y_i < LCL(k)\} \tag{5.21}$$

Therefore, the *ARL* which is the expected value of the run length R is

$$ARL = 1/[1 - P_{inc} + P_{inc}^k (1 - P_{inc})] \tag{5.22}$$

where P_{inc} is the probability that the run has a length within the limits.

More details and a comparison with the original *CCC* chart can be found in Kuralmani *et al.* (2002). In general, the *CCC* approach gives wider control limits while the conditional procedure limits will be narrower, and the conditional procedure provides a warning more often. The procedure is also more sensitive to moderate to large process changes

5.3.3. Optimum conditional control limits

Section 5.1 presented an optimality procedure by which the control limits could be redesigned so as to achieve maximum *ARL* values when the process average is at the nominal level. In the same way, optimum control limits can be developed for the conditional procedure as follows.

For fixed value of k and a prescribed false alarm probability α, when the conditional limits are derived as

$$LCL_{opt} = \frac{\ln[1-(1-\delta)/2]\ln\left\{\dfrac{\ln[1-(1-\delta)/2]}{\ln[(1-\delta)/2]}\right\}}{\ln(1-p_0)\ln\left\{\dfrac{(1-\delta)/2}{[1-(1-\delta)/2]}\right\}} \tag{5.23}$$

and

$$UCL_{opt} = \frac{\ln[(1-\delta)/2]\ln\left\{\dfrac{\ln[1-(1-\delta)/2]}{\ln[(1-\delta)/2]}\right\}}{\ln(1-p_0)\ln\left[\dfrac{(1-\delta)/2}{[1-(1-\delta)/2]}\right]} \tag{5.24}$$

then the average run length has its maximum at $p = p_0$, the prescribed process level.

Example 5.3:

As an example, assume that p=50ppm and the acceptable false alarm probability is 0.05, the value of δ for a desired value of k=5 is 0.88820. Then the control limits of the conditional chart are found to be 1150 and 57682. Compared with the traditional limits, which give 506 and 73776, it can be seen that the control limits for the conditional procedure is significantly narrower. However, because of the additional requirement that an alarm will not be raised if the previous 5 runs are in control, the false alarm rate remains at 0.05.

□

The conditional decision procedure could be compared well with the regular *CCC* chart. Note that the probability of no alarm is given by

$$\beta_{cond} = P_{inc} + P_{inc}^{k}(1 - P_{inc}) \tag{5.25}$$

where, as before, P_{inc} is the probability that the run has a length within the limits.

It has been observed that the operating characteristic curve shows more discrimination in extreme quality levels than the traditional one. This means that the process acceptance has been substantially reduced during the process deterioration or process improvement period. This indicates that the conditional chart is more sensitive over the traditional chart due to the tight control limits. The narrow control limits are very useful in situations like high quality processes in order to detect small process shift in such an environment. The average run length of an optimum conditional chart can be shown to be superior to the traditional chart.

Example 5.4:

From the example 5.3, the optimum limits are 1593 and 79888 while the original conditional limits are 1150 and 57682. Using these limits, the *ARL* curves are drawn in Figure 5.5. The curve is almost shifted towards to the left in order to achieve maximum *ARL* at the desired process average.

The *ARL* is smaller either when the process average is improved or when it has deteriorated. The maximum of the average run length peaks at just p=50ppm. Hence, an alarm will be raised faster no matter which direction the process shifts towards from this value.

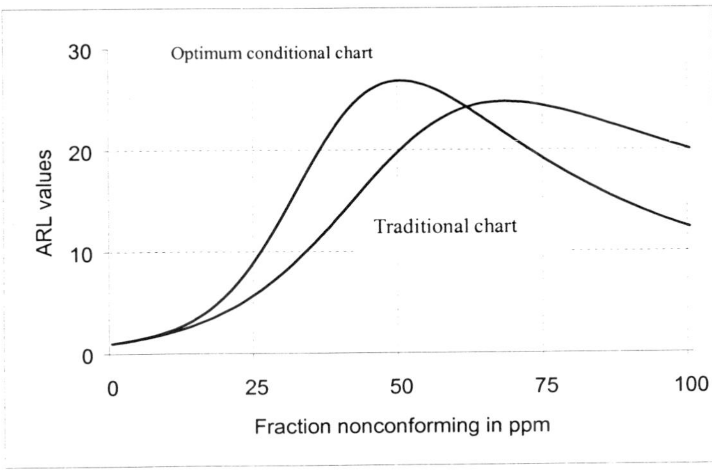

Figure 5.5. *ARL* curves of the optimum conditional and traditional charts for 50ppm.

The optimum control limits can be obtained as described before. Some optimum control limits are calculated for different values of process averages and false alarm probabilities as shown in Table 5.9.

Table 5.9. Values of *LCL and UCL* for optimum conditional charts for various process averages with α=0.0027

k	Limits	Process average					
		0.00001	0.00002	0.00005	0.00010	0.00020	0.00050
1	*LCL*	3585	1793	717	359	179	72
	UCL	496698	248348	99338	49668	24833	9932
2	*LCL*	2529	1265	506	253	126	51
	UCL	538632	269314	107724	53861	26929	10770
3	*LCL*	2067	1034	413	207	103	41
	UCL	562722	281359	112542	56270	28133	11252
4	*LCL*	1789	894	358	179	89	36
	UCL	579908	289953	115979	57988	28993	11595
5	*LCL*	1602	801	320	160	80	32
	UCL	592990	296494	118596	59296	29647	11857
6	*LCL*	1464	732	293	146	73	29
	UCL	603642	301819	120726	60361	30179	12070
8	*LCL*	1271	636	254	127	64	25
	UCL	620263	310130	124050	62024	31010	12402
10	*LCL*	1139	570	228	114	57	23
	UCL	633166	316581	126631	63314	31655	12660
15	*LCL*	935	468	187	94	47	19
	UCL	656257	328127	131249	65623	32810	13122
20	*LCL*	815	407	163	81	41	16
	UCL	672387	336192	134475	67236	33616	13444

5.4. Effects of inspection errors

Most of the results concerning control charts assume that the process parameters are already known. However, they usually have to be estimated from the sampled items through some initial inspection. The inspection error due to misclassification or unclear definition could have significant implication of the results, especially on attribute data (Case, 1980 and Suich, 1988). In Lu *et al.* (2000) this problem is investigated and the main results are summarized here.

5.4.1. Errors involved in inspection activities

An inspection operation may result in two types of inspection errors: classification of a conforming item as nonconforming and classification of a nonconforming item as conforming. These are discussed in Ryan (1989) and Johnson *et al.* (1991). In practice, the process fraction nonconforming p has to be estimated. This estimate might deviate from its true value because of the errors involved in the inspection of items.

Burke *et al.* (1995) gives the relationship between the true and observed process fraction nonconforming as:

$$p_t = (p_o - \theta)/(1 - \theta - \psi) \tag{5.26}$$

where p_o and p_t represent the observed (estimated) p value and the true p value respectively, while θ and ψ denote, respectively, the probability of classifying a conforming item as nonconforming and the probability of classifying a nonconforming item as conforming.

It is clear that for a fixed ψ value, p_t approaches zero when inspection error θ approaches p_o. In practice, this means that for a high quality processes with p_t in the order of 10^{-6} or 10^{-9}, p_o can be as large as θ because of inspection errors.

5.4.2. False alarm probability and average run length

Denote by Y_o the observed cumulative count of conforming items between two successive nonconforming ones. Note that Y_o is the value plotted on the *CCC* chart when there are inspection errors. Again, Y_o follows a geometric distribution with parameter p_o given by

$$\begin{aligned} P(Y_o = n) &= (1 - p_o)^{n-1} p_o \\ &= [1 - (1-\psi)p_t - \theta(1 - p_t)]^{n-1}[(1-\psi)p_t + \theta(1 - p_t)] \end{aligned} \tag{5.27}$$

Usually both inspection errors θ and ψ have to be estimated, and Burke *et al.* (1995) discussed some methods for their estimation. We assume here that the inspection errors are known or already estimated.

With the control limits calculated with an estimated process fraction nonconforming, the actual false alarm probability of the *CCC* chart in the presence of the inspection errors, α_{actual}, is:

$$\begin{aligned}\alpha_{\text{actual}} &= \sum_{i=1}^{LCL} P(Y_o = i) + 1 - \sum_{i=1}^{UCL} P(Y_o = i) \\ &= 1 - (1-p_o)^{LCL} + (1-p_o)^{UCL}\end{aligned} \tag{5.28}$$

Note that this can be rewritten as:

$$p_o = p_t - p_t\psi + (1-p_t)\theta \tag{5.29}$$

and hence, the inspection error θ tends to increase p_o while inspection error ψ tends to reduce p_o. In addition, the inspection error θ affects p_o more significantly than the inspection error ψ does, as the value of $(1-p_t)$ is commonly larger than that of p_t.

We can also investigate the effects of inspection errors on the *CCC* chart when a process change has occurred, based on the average item run length (*AIRL*) values. *AIRL* is an extension of the traditional *ARL* concept since each point on a *CCC* chart corresponds to a different number of items inspected. This quantity is also known as Average Number Inspected (*ANI*). It could be of practical interest to know the average number of items inspected before an alarm rather than the number of points on the *CCC* chart itself.

As the observed cumulative count of conforming items between successive nonconforming items, Y_o is independent on the number of points plotted, the *AIRL* for the *CCC* chart under inspection errors, $AIRL(p_t, \theta, \psi)$, can be shown to be

$$AIRL(p_t,\theta,\psi) = E(Y_o)\times[\frac{1}{1-P(LCL<Y_o<UCL)}]$$
$$= \frac{1}{p_o[1-(1-p_o)^{LCL}+(1-p_o)^{UCL}]} \tag{5.30}$$

where p_o, LCL and UCL are the same as those in the previous sections.

Example 5.5:

Assume that in an automated manufacturing process, the true process fraction nonconforming is $p_t = 0.001$ and the desired false alarm probability α_{desired} takes the standard value of 0.0027. Then the control limits for the corresponding *CCC* chart are $UCL = 6604.35$; and $LCL = 1.35$

With inspection errors included in consideration, the *AIRL* for the *CCC* chart is

$$AIRL(0.001,\theta,\psi) = \frac{1}{(0.001-0.001\psi+0.999\theta)}\cdot$$
$$\frac{1}{[1-(0.999+0.001\psi+0.999\theta)^{1.35}+(0.999+0.001\psi+0.999\theta)^{6604.35}]}$$

Four different combinations of inspection errors are considered: $(\theta,\psi)=(0,0)$, (0.0001, 0), (0, 0.2) and (0.0001, 0.2). The first case corresponds to the inspection without errors while the other three represent different error rates. More details have been shown in Lu (1999) and are omitted here.

The numerical results indicate that the inspection error combinations $(\theta,\psi)=(0.0001,0)$ and $(\theta,\psi)=(0,0.2)$ shift, respectively, the *AIRL* curve to the left and to the right of the one obtained under error-free inspection. At the same time, the inspection error combination $(\theta,\psi)=(0.0001,0.2)$ shifts the *AIRL* curve to the left of the *ANI* curve for $(\theta,\psi)=(0,0.2)$ and to the right of the *ANI* curve for

$(\theta,\psi)=(0.0001,0)$. The explanations for these observations are straightforward based on theoretical discussions.

Another interesting result is that the *AIRL* curve for (0.0001,0.2) is closer to the *AIRL* curve under error-free inspection than the *AIRL* curves for (0.0001,0) and (0,0.2). The reason is that θ error tends to increase, while ψ error tends to decrease, the observed process fraction nonconforming and thus their individual effects on the *AIRL* curves are partially offset by each other. This implies that it might be better to have a balanced approach to reducing inspection errors rather than try to reduce one type of inspection error at the expense of increasing the other type of inspection error.

□

5.4.3. Adjusted control limits under inspection errors

When inspection errors are present, it is sometimes possible to modify the control limits to reduce the effect of inspection errors. With the adjusted control limits, a control chart can provide an *ARL* curve that is closer to the one under error-free inspection.

A study of the *AIRL* curves shows that the *AIRL* values with inspection errors differ significantly from the one under error-free inspection, when the process is in control. Therefore, one possible way is to adjust the control limits so that they can provide the same in-control *AIRL* values as those under error-free inspection. Then we can proceed to check whether the *AIRL* curves also "match" when a process shift has occurred. Let UCL_a, CL_a, LCL_a, $\alpha^*_{\text{desired}}$ and $AIRL_a$ denote the adjusted upper control limit, central line, lower control limit, desired false alarm probability and average number inspected respectively for the *CCC* chart in the presence of inspection errors.

Lu *et al.* (2000) obtained the adjusted control limits expressed as

$$\begin{aligned} UCL_a &= \ln(\alpha^*_{\text{desired}}/2)/\ln(1-p_o) \\ CL_a &= \ln(0.5)/\ln(1-p_o) \\ LCL_a &= \ln(1-\alpha^*_{\text{desired}}/2)/\ln(1-p_o) \end{aligned} \tag{5.31}$$

The *AIRL* provided by the adjusted control limits is

$$AIRL_a = \frac{1}{p_o[1-(1-p_o)^{LCL_a}+(1-p_o)^{UCL_a}]} \tag{5.32}$$

On the other hand, the in-control value of the *AIRL* under error-free inspection is given by

$$AIRL_{\text{in-control}}(p_t,\theta,\psi) = \frac{1}{\alpha_{\text{desired}}\,p_t} \tag{5.33}$$

By equating these two, we have

$$\alpha^*_{\text{desired}} = \frac{\alpha_{\text{desired}}\,p_t}{p_o} = \frac{\alpha_{\text{desired}}\,p_t}{[(1-\psi)p_t+(1-p_t)\theta]} \tag{5.34}$$

Hence, the adjusted control limits are:

$$UCL_a = \frac{\ln\left\{\dfrac{0.5\alpha_{\text{desired}}\,p_t}{[(1-\psi)p_t+(1-p_t)\theta}\right\}}{\ln[1-(1-\psi)p_t-(1-p_t)\theta]} \tag{5.35}$$

$$CL_a = \frac{\ln(0.5)}{\ln[1-(1-\psi)p_t-(1-p_t)\theta]} \tag{5.36}$$

and

$$LCL_a = \frac{\ln\left\{1-\dfrac{0.5\alpha_{\text{desired}}\,p_t}{[(1-\psi)p_t+(1-p_t)\theta}\right\}}{\ln[1-(1-\psi)p_t-(1-p_t)\theta]} \tag{5.37}$$

Example 5.6:
As in the previous example, let p_t be 0.001 and α_{desired} be 0.0027. The adjusted control limits for the inspection error combination $(\theta,\psi)=(0.0001,0.2)$ are

$$LCL_a = 1.7;\ CL_a = 769.8;\ UCL_a = 7221.5$$

The adjusted average number inspected, $AIRL_a$, is

$$AIRL_a = \frac{1}{p_o}\cdot\frac{1}{[1-(1-p_o)^{1.7}+(1-p_o)^{7221.5}]}$$

The adjusted control limits for all the four inspection error combinations are calculated in the same way and shown in Table 5.10.

□

Table 5.10. Some numerical values of the control limits under inspection errors

(θ,ψ)	CL_a	CL_a	UCL_a
(0,0)	1.4	692.8	6604.3
(0.0001,0)	1.1	629.8	6090.2
(0,0.2)	2.1	866.1	7977.4
(0.0001,0.2)	1.7	769.8	7221.5

5.5. Effects of estimated control limits

In implementing control charts in practice, process parameters are often unknown and have to be estimated. This is an essential step in setting up control charts. The estimation is usually done by collecting initial samples from a stable running process and estimating the control limits based on the inspection results of these samples. An important question is, when the sample size is small, what is the effect on the control limits, see, e.g., Chen (1997) and Braun (1999).

This question is essential in implementing the *CCC* chart as few items are nonconforming in this type of production processes. This makes an accurate estimation of the process fraction nonconforming p very difficult. The fluctuations in estimated p values will possibly make the estimated control limits be far from the true limits, as the control limits are a logarithmic-type function of p. Under such circumstances, the misleading results indicated by the *CCC* charts may lead to wrong management decisions.

Let Y_i, $i = 1,...,n$, be the count the number of the conforming items between the (i-1):th and i:th nonconforming item from a stable process running in an automated manufacturing environment. Then, if the process fraction nonconforming p is a known constant, count Y_i for $i = 1,...,n$ will be plotted on the *CCC* chart with the control limits:

$$UCL = \ln(\alpha_{\text{desired}} / 2) / \ln(1-p) \tag{5.38}$$

and

$$LCL = \ln(1-\alpha_{\text{desired}} / 2) / \ln(1-p) \tag{5.39}$$

where α_{desired} is the desired false alarm probability of the *CCC* chart. Note that α_{desired} can be very different from the true value when the fraction nonconforming p is estimated and the true value is the focus of the discussion here.

Let A_i denote the event that variable Y_i either exceeds *UCL* or is less than *LCL*. Then the probability of event A_i, $P(A_i)$, is

$$P(A_i) = P(Y_i > UCL \text{ or } Y_i < LCL) = 0.0027 \tag{5.40}$$

if α_{desired} is 0.0027.

Let m be the number of the sampled items and d denote the number of nonconforming ones among these m sampled items. If p is unknown, it can be estimated as

$$\overline{p} = d / m \tag{5.41}$$

Correspondingly, the estimated limits for the *CCC* chart, $\overline{UCL}$ and $\overline{LCL}$, are

$$\overline{UCL} = \ln(\alpha_{\text{desired}} / 2) / \ln(1 - \overline{p}) \tag{5.42}$$

and

$$\overline{LCL} = \ln(1 - \alpha_{\text{desired}} / 2) / \ln(1 - \overline{p}) \tag{5.43}$$

Note that $\overline{p}$ is a logarithmic type function, and the distributions for the estimators $\overline{UCL}$ and $\overline{LCL}$ are not straightforward. However, they can be investigated by Monte Carlo simulations (Lu, 1999).

Let B_i denote the event that count Y_i either exceeds $\overline{UCL}$ or is less than $\overline{LCL}$. Then the probability of even B_i, $P(B_i)$, can be written as

$$P(B_i) = P(Y_i > \overline{UCL} \text{ or } Y_i < \overline{LCL}) \tag{5.44}$$

This probability is called the actual false alarm probability of the *CCC* chart, α_{actual}, when p is unknown. Since the count Y_i follows a geometric distribution depending solely on p while estimators $\overline{UCL}$ and $\overline{LCL}$ follow the distributions depending solely on p and m, the α_{actual} will also depend solely on p and m.

Using the concept of conditional probability, we have

$$\begin{aligned} P(B_i) &= \sum_{d=0}^{m} P(B_i \mid N = d) P(N = d) \\ &= \sum_{d=0}^{m} P(B_i \mid N = d) \binom{m}{d} p_0^d (1 - p_0)^{m-d} \end{aligned} \tag{5.45}$$

where

$$P(B_i \mid N = d)$$
$$= P\{X_i > \overline{UCL}(N) \mid N = d\} + P\{X_i < \overline{LCL}(N) \mid N = d\}$$
$$= (1-p)^{\ln(\alpha/2)/\ln(1-d/m)} - (1-p)^{\ln(1-\alpha/2)/\ln(1-d/m)} + 1$$

Using a similar argument, the average run length (*ARL*) in terms of the number of points plotted to an alarm, and the standard deviation of the run length (*SDRL*) can be seen to have the forms

$$ARL(p_0, p) = E_N[1/\alpha(N)]$$

and

$$SDRL(p_0, p) = \sqrt{Var_N[1/\alpha(N)] + E_N[(1-\alpha(N))/\alpha^2(N)]}$$

They can be computed as

$$E_N[1/\alpha(N)] = \sum_{d=0}^{m} \frac{1}{\alpha(d)} \binom{m}{d} p_0^d (1-p_0)^{m-d} \tag{5.46}$$

and

$$E_N[1/\alpha^2(N)] = \sum_{d=0}^{m} \frac{1}{\alpha^2(d)} \binom{m}{d} p_0^d (1-p_0)^{m-d} \tag{5.47}$$

Recently, Yang *et al.* (2002) investigated the false alarm probability and average run length for the case of estimation errors. Extensive simulation results are presented. Readers are referred to that paper for more technical details.

CHAPTER 6

Some Extensions to the Geometric Model

Although the simple geometric model is useful for high-quality process monitoring, some extensions are possible for specific purposes. Chapter 5 dealt with the modification of the geometric chart from an implementation point of view. This chapter presents some further extensions of the geometric model for process modeling and analysis.

Instead of monitoring the number of items inspected before a nonconforming one is observed, we could monitor the process by using the number of items inspected before r nonconforming items are observed. It is well-known that this number will follow the negative binomial distribution which is a generalization of the geometric distribution. A chart on this basis, called *CCC-r* chart, is described in Section 6.1. Section 6.2 presents a case when the number of nonconformities, or defects, can be counted, but there are a large number of items with no defects.

A modified Poisson model, or the zero-inflated Poisson model, is recommended to model the count data and statistical procedure based on this model is discussed. A dual control charting procedure using the *CCC* chart and *c*-chart is presented for effective process monitoring. Sections 6.3 and 6.4 present some models and results for the case when data are correlated, a situation that is common when items are produced and automatically checked individually in a continuous production environment.

6.1. Negative binomial control chart

The idea of *CCC* chart can be extended to process monitoring by considering the number of items inspected until a fixed number of nonconforming items are observed. We denote the control chart based on this count of items inspected until r nonconforming items are observed by *CCC-r* chart. The idea of *CCC-r* control chart is presented in Xie *et al.* (1999) and further discussed in Sun and Zhang (2000) and Ohta (2001). We briefly describe this approach here.

6.1.1. Control limits with negative binomial distribution

Let p be the probability that an item is nonconforming. The items are inspected until r nonconforming ones are counted. Let X be the cumulative count of items inspected until the observation of r nonconforming ones. It is well known that count X follows the negative binomial distribution as described in Section 1.4. The probability distribution function of X is given by

$$P\{X=i\}=\binom{i-1}{r-1}p^r(1-p)^{i-r},\ i=r,r+1,... \tag{6.1}$$

which is the negative binomial distribution with parameters (r, p).

Note that a geometric random variable is a special case of negative binomial with parameters $(1, p)$. The negative binomial random variable can also be seen as the sum of r independent identical geometric random variables with parameters $(1, p)$. Hence the *CCC-r* chart presented is a generalization of the *CCC* chart.

The cumulative distribution function of negative binomial count *X* can be written as

$$F(n,r,p) = P\{X \le n\} = \sum_{i=r}^{n} \binom{i-1}{r-1} p^r (1-p)^{i-r} \tag{6.2}$$

When *r* is equal to one, as the case of *CCC* chart, the distribution reduces to that of geometrically distributed counts. When *r* is two, we have

$$\begin{aligned} F(n,2,p) &= p^2 \left[\frac{1-(n-1)(1-p)^n - n(1-p)^{n-1}}{p^2} \right] \\ &= 1 + (n-1)(1-p)^n - n(1-p)^{n-1} \end{aligned} \tag{6.3}$$

When the parameter $r = 3$, the negative binomial distribution has more complicated forms. However, it can be easily calculated with a standard computer program.

A process can be monitored based on *CCC-r* count. The probability limits can be derived in a standard way. For an acceptable risk of false alarm, α, the upper control limit *UCL* and the lower control limit *LCL* of the *CCC-r* chart can be obtained as the solutions of the following equations,

$$F(UCL,r,p) = \sum_{i=r}^{UCL} \binom{i-1}{r-1} p^r (1-p)^{i-r} = 1 - \frac{\alpha}{2} \tag{6.4}$$

and

$$F(LCL,r,p) = \sum_{i=r}^{LCL} \binom{i-1}{r-1} p^r (1-p)^{i-r} = \frac{\alpha}{2} \tag{6.5}$$

while the center line, *CL*, is given as the solution of

$$F(CL,r,p) = \sum_{i=r}^{CL} \binom{i-1}{r-1} p^r (1-p)^{i-r} = \frac{1}{2} \tag{6.6}$$

The above equations can be solved numerically using standard computer programs. Some numerical values for *CCC*-1, *CCC*-2 and *CCC*-3 charts are shown in Table 6.1.

Table 6.1. Some control limits of *CCC*-1, *CCC*-2 and *CCC*-3 charts, with $\alpha = 0.0027$

p	*CCC*-1 chart			*CCC*-2 chart			*CCC*-3 chart		
	UCL	*CL*	*LCL*	*UCL*	*CL*	*LCL*	*UCL*	*CL*	*LCL*
.0001	66074	6932	13	91895	16786	530	114341	26746	2118
.0002	33035	3466	7	43522	8391	265	52817	13368	1060
.0003	22023	2311	5	29487	5594	177	35959	8913	707
.0004	16516	1733	4	22316	4196	133	27284	6685	531
.0005	13212	1386	3	17953	3357	107	21990	5349	425
.0006	11010	1155	3	14744	2797	89	17973	4457	354
.0007	9437	990	2	12699	2398	77	15505	3820	304
.0008	8257	867	2	11158	2098	67	13640	3343	266
.0009	7339	770	2	9831	1865	60	11982	2971	237
.0010	6605	693	2	8876	1678	54	10833	2674	213
.0020	3301	347	1	4436	839	27	5414	1337	107
.0040	1649	173	1	2222	420	14	2715	669	54
.0060	1098	116	1	1479	280	10	1806	446	37
.0080	823	87	1	1110	210	8	1355	334	28
.0090	731	77	1	986	187	7	1204	297	25
.0100	658	69	1	886	168	6	1083	268	23
.0200	328	35	1	441	84	4	539	134	12
.0400	162	17	1	219	42	2	268	67	7
.0500	129	14	1	175	34	2	213	54	6

6.1.2. Some implementation issues

For the *CCC-r* chart, the selection of r can be considered to be a subjective issue and this is similar to the case of sample size in traditional Shewhart charts. On the other hand, usually r should be a small number and the selection is easier. The control limits of the *CCC-r* chart increase very quickly when process fraction nonconforming p approaches zero. The control limits of the *CCC-r* charts with a large r are much larger than those with small r for the same p level. Usually, an r value of 2-5 should be sufficient in practice.

When the p value is small enough, the center line which indicates the average number of items inspected until a point is plotted is very large. It is then not appropriate to use the *CCC-r* chart with a large r value. Hence for $p<0.0001$, it is advisable to use the *CCC* chart which has been studied in Chapter 3. The traditional *CCC* chart is the same as our *CCC*-1 chart here.

In order to apply the *CCC-r* chart, the target value of the process fraction nonconforming, p_0, has to be assumed known or estimated. Let N denote the number of all the items produced and d the number of nonconforming items detected in the N items produced. Then p_0 can be estimated as

$$\bar{p}_0 = \frac{d}{N} \tag{6.7}$$

It should be mentioned that the decision-making procedure here is different from that of the traditional p- or np-chart. On the *CCC-r* control chart, if a point is above the upper control limit, the process is believed to have improved; while if a point is below the lower control limit, the process is considered to have deteriorated. This is opposite to the situation in the p- chart or np-chart. Hence, the *CCC-r* chart has the advantage of giving the chart user a chance to detect process improvement in an easy way, while the traditional p- or np-chart usually cannot do so.

6.1.3. Some analytical results

The average run length usually denotes the average number of samples taken before a process change is detected. It is also equivalent to the

It can be noted that the *AIRL* increases quickly when $p < 0.0002$. The *AIRL* is the product of the average number of points plotted on the chart up to an alarm signal, $1/(1-\beta_r)$, and the average number of items produced when a point is plotted, r/p. When process improves, the former tends to decrease while the latter tends to increase the value of *AIRL*.

Moreover, it can be noted that the effect of the latter on *AIRL* becomes more significant compared to that of the former as p approaches zero. Therefore, their combined effects lead to the observation that *AIRL* decreases first and then increases when the process improves.

□

6.2. Modified Poisson model

For controlling the number of nonconformities in a sample, the *c*-chart is widely used. *c*-charts are based on the assumption that the number of nonconformities in a sample is Poisson distributed with parameter *c*. The Poisson assumption is a good approximation in practice, although in near zero-defect (ZD) environment this may not be true, particularly when a process is randomly affected by mechanisms, which lead to a large number of nonconformities.

In this section, we present the use of a modified Poisson model for this type of processes. The formulation is first given followed by a physical interpretation of the model in an earlier paper, with the procedure reported in Xie *et al.* (1995). This idea further investigated in Chang and Gan (1999). A Shewhart-like control chart is used here. A combined control chart for which the *CCC* chart is an integral part is then presented for better controlling both the parameters in the process model.

6.2.1. A model for zero-defect process subject to random shock

Suppose that for a process, the individual product units are basically perfect, but there exists a random mechanism, referred to as shocks here, which leads to a number of nonconformities in the sample whenever it is activated. The shocks occur randomly with a low probability so that a large number of zero-nonconformity samples are observed; however, when a sample is affected by a shock, it can contain any number of nonconformities. We call the model one of a ZD process subject to random shocks.

The model is applicable in many practical situations in which some kind of unidentified and uncontrollable mechanism takes place to cause nonconformities in an otherwise near perfect process. The proposed model is useful in developing a procedure for determining whether an out-of-control situation exists.

The model is a two-stage model for the analysis of nonconformity data with many zeros and can be formulated as in the following. In this model, each sample is treated as a unit and each unit contains a number of discrete components. For each unit, we assume that

(a) there is a probability, p, that the unit is a nonconforming (defective) one containing some nonconformities, i.e., nonconforming components. Therefore, $1-p$ is the probability that the unit is perfect;

(b) in a nonconforming unit, the number of nonconformities in the unit follows a certain probability distribution.

In an ideal situation, only perfect units are observed. However, a unit can be affected by a random mechanism that introduces nonconformities. If such a mechanism (which may or may not be eventually controllable) occurs, there will be some nonconformities in the unit.

It can be easily shown that

$$P(\text{no nonconformities in the sample}) = (1-p) + pe^{-\lambda} \tag{6.10}$$

and

$$P(k \text{ nonconformities in the sample}) = p\frac{\lambda^k e^{-\lambda}}{k!}, \quad k = 1,2,... \tag{6.11}$$

The distribution of the number of nonconformities can be treated as a midified Poisson distribution. Note that the probability of no nonconformities can be high dependent on the value of p, the reason behind the strings of zero-nonconformity samples. This model is also known as zero-inflated Poisson distribution, as the same phenomenon has been observed in other fields (Bohning, 1998).

Estimation of the parameters

In this model, p is the probability of occurrence of random shock and λ is the average number of nonconformities in a sample when a shock manifests itself. Given a data set $\{n_k, k=0,1,2,...,n\}$ where n_k is the number of samples containing k nonconformities and n the total number of samples in the study, we can easily estimate p and λ by the method of maximum likelihood.

The likelihood function of p and λ for the data set $\{n_k, k=0,1,2,...,n\}$ is given by

$$L(p,\lambda) = \prod p_k^{n_k} = \{(1-p) + pe^{-\lambda}\}^{n_0} \prod \left\{p\frac{\lambda^k e^{-\lambda}}{k!}\right\}^{n_k} \tag{6.12}$$

Taking the logarithm of this likelihood function, it can be shown that the maximum likelihood estimates of p and λ can be determined by solving the following likelihood equations,

$$\frac{n_0(-1+e^{-\lambda})}{(1-p)+pe^{-\lambda}}+\frac{n_d}{p}=0 \tag{6.13}$$

and

$$\frac{-n_0 pe^{-\lambda}}{(1-p)+pe^{-\lambda}}+\frac{m}{\lambda}-n_d=0 \tag{6.14}$$

where $n_d = n - n_0$ is the total number of units in which one or more nonconformities are found and

$$m=\sum_{k=1}^{\infty} kn_k$$

is the total number of nonconformities found.

These likelihood equations have to be solved numerically. Note that only three quantities, n_d, m and n, need to be recorded for the estimation.

6.2.2. Control limits based on modified Poisson model

As for other statistical process control procedures, we now assume that the model parameters are known or accurately estimated using the existing data. They could be determined, e.g. by maximizing the likelihood equations. Let n_u denote the upper control limit for a control chart based on the number of nonconformities per sampled unit. Then, for an acceptable false alarm probability of α, its value can be determined by

$$P(n_u \text{ or more nonconformities in a sample}) \le \alpha$$

It can be noted that there can be two out-of-control states, which are not distinguished by the approach used here. If a large number of nonconformities are observed, this may be caused by an increase in either λ or p.

If λ is increased, then it is certain that the number of nonconformities in a sample will increase. If p is increased, then it will be more often that none-zero nonconformity samples will be observed which also lead to samples with a large number of nonconformities. Furthermore, if λ is decreased and p is increased or vise versa, then more sophisticated procedures are needed to identify yet another out-of-control state, which will be discussed later.

Suppose that $\alpha = 0.001$, the decision criteria proposed for the process is equivalent to

$$\sum_{k=n_u}^{\infty} p \frac{\lambda^k e^{-\lambda}}{k!} \le 0.001$$

or

$$(1-p) + pe^{\lambda} + \sum_{k=1}^{n_u - 1} p \frac{\lambda^k e^{-\lambda}}{k!} > 0.999 \tag{6.15}$$

Note that n_u depends on both p and λ. However, it can be easily solved because it takes only discrete values.

Example 6.2:

As an example, a set of data on read-write errors discovered in a computer hard disk which can exhibit a string of zeros before a large count of errors suddenly appears, is used. Table 6.3 shows a set of actual data of this type (Xie and Goh, 1992).

Using the model proposed in this paper, we have n=208, m=242 and n_d=28. The maximum likelihood estimates of λ and p are obtained, after solving the likelihood equations, as 8.64 and 0.135 respectively. The upper control limit is then determined to be 17 (inclusive). A plot showing the traditional c-chart and compared with the new limit is displayed in Figure 6.1.

Table 6.3. A data set of nonconformities in a near ZD environment (number of nonconformities; read from left to right and down)

0	0	0	0	0	0	0	0	0	0	0	0	1	0	0
0	0	6	0	9	11	0	1	2	0	0	0	0	0	0
0	0	3	3	0	0	5	0	15	6	0	0	0	4	2
0	0	0	1	1	0	1	0	0	0	0	0	0	0	0
0	0	0	0	0	0	0	0	0	0	0	0	0	0	0
0	0	0	0	0	75	0	0	0	0	75	0	0	0	0
0	0	0	0	0	0	0	0	0	0	0	0	0	2	0
0	0	0	0	0	0	0	0	0	0	0	0	1	0	0
0	0	0	0	1	0	0	0	0	0	0	0	0	0	0
0	0	0	0	0	0	1	0	0	1	0	0	0	0	0
0	0	0	0	0	0	0	0	9	0	0	2	0	0	0
0	0	0	0	0	0	0	1	0	0	0	0	0	0	0
0	0	0	0	0	0	0	0	0	0	0	0	0	0	0
0	0	0	2	0	0	0	1	0	0	0	0	0		

From Figure 6.1, only two points are "out-of-control" with sufficient certainty. Suppose that they are removed after the assignable causes are detected, we can, as in the traditional way, get a revised control limit by using n=206, m=92 and n_d=26. The revised process characteristics λ and p are 5.4 and 0.132 respectively. The upper control limit is then 8. It may then be suspected that three remaining points are due to assignable causes.

□

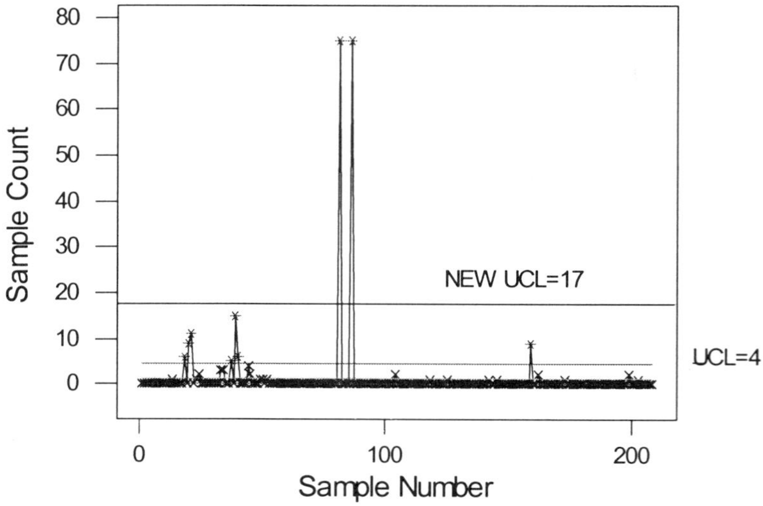

Figure 6.1. The chart with the new model compared with the traditional *c*-chart (with *UCL*=4) for which many points are out-of-control.

6.2.3 Combined control charts for modified Poisson model

With the revised control limits, the proposed model still suffers a few drawbacks as the conventional Shewhart chart in this case, although the probability of false alarm is considerably reduced. Process improvement cannot be detected without further run rules and there is still much unnecessary plotting. Furthermore, when the chart signals an out-of-control alarm, we cannot tell which of the model parameters have changed.

An interesting approach is adopted in Xie *et al.* (1995). Based on this and motivated by the idea of X-bar and R-charts, two control charts can be used to monitor this type of process.

We can count the number of conforming items until a nonconforming one is detected. We record the count and also the number of nonconformities in the nonconforming item. A *CCC* chart is used to monitor the count of the consecutive conforming items, and this indicates the frequency of the random shock. We also use a c-chart to monitor the number of nonconformities contained in a nonconforming item, which is related to the severity of the random shock.

It is clear that the nonconforming items appear at a rate of

$$p_0 = p(1 - e^{-\lambda})$$

where p is the frequency of the random shock. When the severity of the random shock is unchanged, the probability of getting a nonconforming unit only depends on p. A standard *CCC* chart can be set up to monitor this. For an acceptable risk of false alarm, α, the control limits are as before,

$$UCL_p = \frac{\ln(\alpha_p / 2)}{\ln(1 - p_0)} \tag{6.16}$$

and

$$LCL_p = \frac{\ln(1 - \alpha_p / 2)}{\ln(1 - p_0)} \tag{6.17}$$

where α_p is the actual false alarm probability. For the number of nonconformities in a nonconforming item, we apply the traditional c-chart. This can be done because

$$P(k \text{ nonconformities in an item} \mid \text{a random shock occured})$$
$$= \frac{P(k \text{ nonconformities in an item AND a random shock occured})}{P(\text{random shock occurs})}$$

which is

$$\sum_{k=n_u}^{\infty} \frac{\lambda^k e^{-\lambda}}{k!}, k = 0,1,2,... \tag{6.18}$$

This is basically a Poisson distribution with parameter λ. The traditional c-chart can be used to monitor this count of nonconformities. The control limits are

$$UCL_\lambda = \lambda + 3\sqrt{\lambda}$$

and

$$LCL_\lambda = \lambda - 3\sqrt{\lambda}$$

The charting can be done straightforwardly. A charting procedure is given in Figure 6.2.

The decision making has to be based on both charts jointly. If any of the charts show an out-of-control signal, then the process is likely to have changed. When the frequency of random shock is increased, the *CCC*-chart should show a signal. When the severity of random shock is increased, the c-chart should show a signal. Hence, decision making is actually made easy in the case of an out-of-control signal. We can tell which type of causes we should look for.

Furthermore, sincc thc *CCC* chart is used here, it enables improvement detection which is an important strength of the *CCC* chart. That is, when the *CCC* chart shows a point above the upper control limit, it could be expected that some assignable causes can be found and it indicates an improved process.

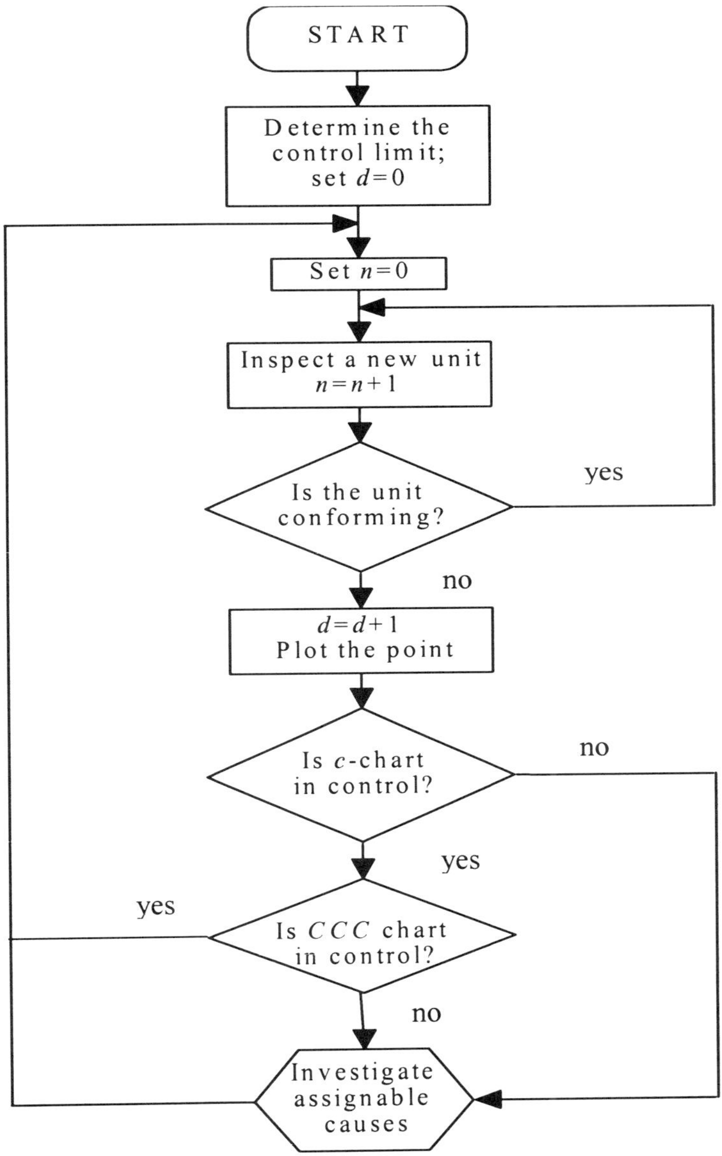

Figure 6. 2. The charting procedure for the combined chart.

6.3. A Markov model for dependent counts

The simple *CCC* chart presented in Chapter 3 is shown to be very useful for the monitoring of high quality processes. However, as for most commonly used Shewhart chart, it is assumed that the items inspected are independent, and this is probably not the case for high quality processes because of short sampling intervals. In the presence of correlation, the conventional *CCC* chart will give a different false alarm probability than what is specified. We first present a study of using a Markov model and in Section 6.4 a model based on exchangeable variables is discussed.

6.3.1. A Markov model for a serially dependent process

Broadbent (1958) first describes a production process of bottles from a mould in an automatic manufacturing scheme. It is reported that the conforming and nonconforming bottles occurred in runs and therefore it is proposed that Markov model be used for the serially dependent production process wherein the probability of finding a nonconforming or conforming unit depends on the status of the immediate preceding unit. Some recent work on Markov-dependent production processes includes McShane and Turnbull (1991) and Vellaisamy and Sankar (2001).

Bhat and Lal (1990) derived upper and lower control limits for attribute control charts when the process possesses Markov dependency, and they relied on the traditional p and np setting. Stimson and Mastrangelo (1996) studied the monitoring of serially dependent processes with attribute data obtained from the multistations of production. Lai *et al.* (2000) further investigated this issue in the context of *CCC* related data.

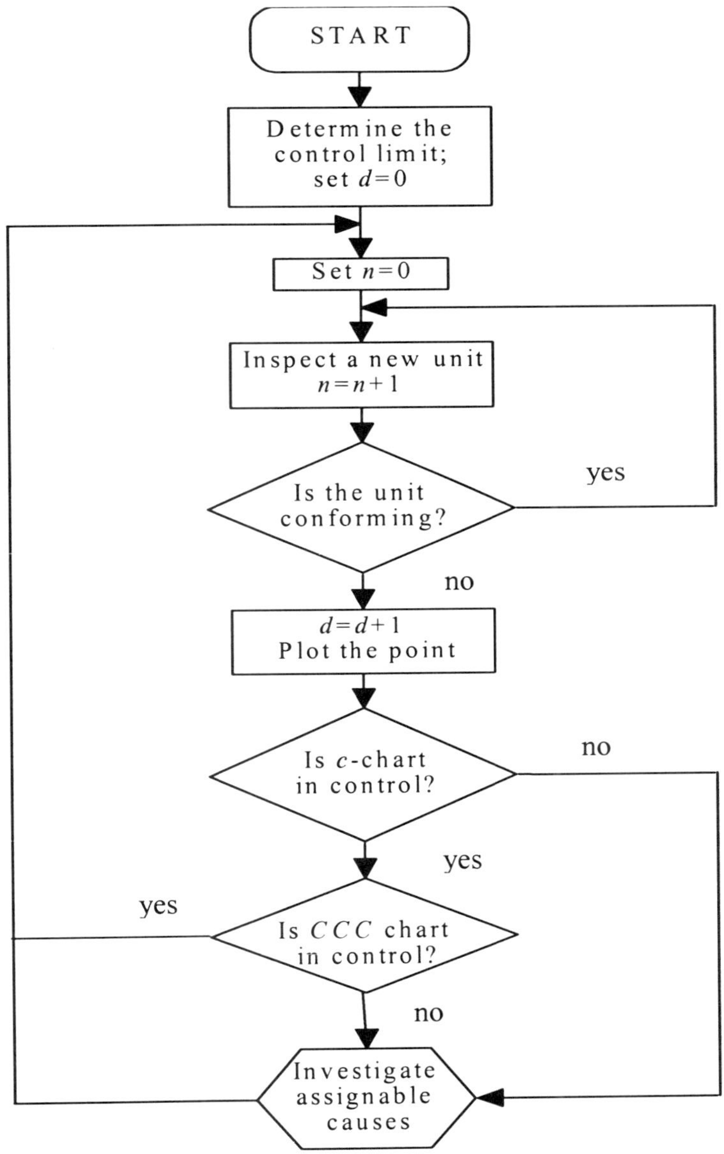

Figure 6. 2. The charting procedure for the combined chart.

Example 6.3:
We applied this combined chart to the data set 1 in Table 6.3. Two charts are shown jointly in Figures 6.3 and 6.4.

Table 6.4. The data of cumulative conforming item and number of nonconformities in nonconforming items as in Table 6.3

Data #1 - samples containing one or more nonconforming items

1	6	9	11	1	2	3	3	5	15	6	4	2	1
1	1	75	75	2	1	1	1	1	9	2	1	2	1

Data #2 - the consecutive number of conforming samples

13	5	2	1	2	1	9	1	3	2	1	4	1	4
1	2	29	5	18	14	7	17	3	14	3	11	26	4

It is clear that in this case, the two items with 75 nonconformities each are out-of-control. It is the *c*-chart that shows this and it is expected that this out-of-control point is probably caused by sever random shock.

Although this could have been guessed from the beginning, an interesting fact is that the *CCC* chart, with control limit computed using the first part of the data would show an out-of-control signals at the nonconforming number 18. It is clear that the number of consecutive conforming items is much larger than before. In this case, the process is probably improved, and this is an improvement signal that could not be detected with any single chart.

□

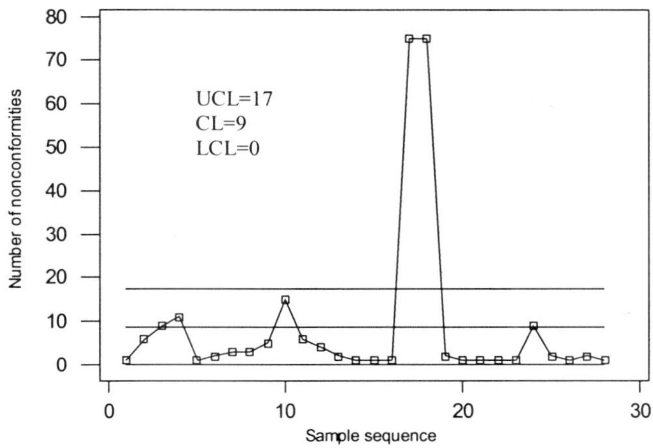

Figure 6.3. The *c*-chart based on data#1.

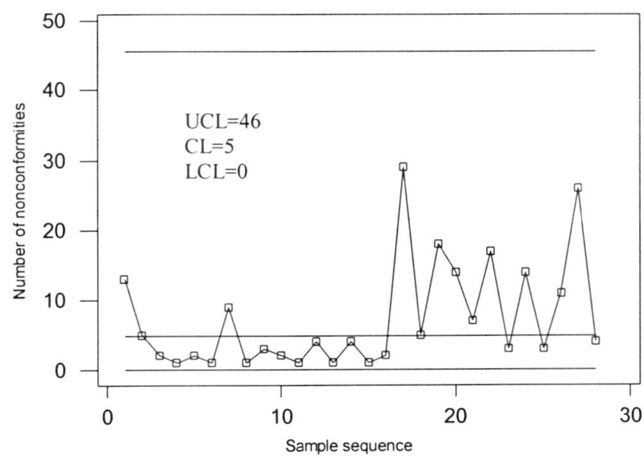

Figure 6.4. The *CCC*-chart based on data#2.

6.3. A Markov model for dependent counts

The simple *CCC* chart presented in Chapter 3 is shown to be very useful for the monitoring of high quality processes. However, as for most commonly used Shewhart chart, it is assumed that the items inspected are independent, and this is probably not the case for high quality processes because of short sampling intervals. In the presence of correlation, the conventional *CCC* chart will give a different false alarm probability than what is specified. We first present a study of using a Markov model and in Section 6.4 a model based on exchangeable variables is discussed.

6.3.1. A Markov model for a serially dependent process

Broadbent (1958) first describes a production process of bottles from a mould in an automatic manufacturing scheme. It is reported that the conforming and nonconforming bottles occurred in runs and therefore it is proposed that Markov model be used for the serially dependent production process wherein the probability of finding a nonconforming or conforming unit depends on the status of the immediate preceding unit. Some recent work on Markov-dependent production processes includes McShane and Turnbull (1991) and Vellaisamy and Sankar (2001).

Bhat and Lal (1990) derived upper and lower control limits for attribute control charts when the process possesses Markov dependency, and they relied on the traditional p and np setting. Stimson and Mastrangelo (1996) studied the monitoring of serially dependent processes with attribute data obtained from the multistations of production. Lai *et al.* (2000) further investigated this issue in the context of *CCC* related data.

In a two-state Markov chain model, one considers two types of probabilities, namely:

a = probability of obtaining a nonconforming unit if the preceding unit is conforming

b =probability of obtaining a conforming unit if the preceding unit is nonconforming.

The process is represented by a two-state Markov chain with transition matrix:

$$P = \begin{matrix} S \\ F \end{matrix}\begin{bmatrix} 1-a & a \\ b & 1-b \end{bmatrix} \tag{6.19}$$

where S (success) stands for finding a conforming unit and F (failure) stands for a nonconforming unit. Bhat and Lal (1990) derived the upper and lower control limits for attribute control charts (i.e. p- chart and np-chart) when the process is under the Markov dependent condition just described. Their control limits were obtained by using the distribution of the number of successes in Markov trials derived by Bhat and Lal (1988). They have not considered the effect of Markov dependence in a high quality improvement.

Under the two states Markov model of the process, the long run fraction nonconforming can be shown to be $p = a/(a+b)$ and the serial correlation is $d = 1-(a+b)$. Since we may assume that the production process is under equilibrium, the Markov dependent production may be identified by two parameters, the fraction defective p and the serial correlation coefficient d. For d=0, the Markov process becomes a sequence of independent Bernoulli processes.

For high-quality processes, the long run fraction nonconforming p, is relatively small. This implies that a is small and b should be relatively large. If the value of a can be ignored in the expression for the serial correlation, we have that

$$d=1-(a+b)\approx 1-b$$

and when b is close to 1, d is close to zero. However, we will see later that even for a small value of d, the serial correlation should not be ignored for the case of high quality process.

When using the Markov model, the parameters will have to be estimated. This can be done in the following way. Let p_{ij} be the transition probability that corresponds to 6.31. That is,

$$p_{00}=1-a,\quad p_{01}=a,\quad p_{10}=b,\quad p_{11}=1-b$$

and let n_{ij} denote the number of times the process moves from state i to state j (i,j takes value 0 or 1) with states 0 and 1 representing conformance and nonconformance of an item respectively. The maximum likelihood estimate of p_{ij} is (Bhat and Lal, 1990)

$$p_{ij}=\frac{n_{ij}}{n_{io}+n_{i1}},\quad i,j=0,1 \tag{6.20}$$

By the relationship between (p,d) and (a,b) and the property of maximum likelihood estimators, we have

$$\hat{p}=\frac{\hat{p}_{01}}{\hat{p}_{01}+\hat{p}_{10}} \text{ and } \hat{d}=1-(\hat{p}_{01}+\hat{p}_{10}) \tag{6.21}$$

6.3.2. The distribution of run length

When the production process is independent, the distribution of X, the number of units observed to obtain a nonconforming unit, is well known. We now assume that the production process of attributes is a Markov chain specified before.

By using conditional probability property, i.e., $P(A/B)=P(A)P(B/A)$ we find that

$$P(Y=0)=1-b=d+p(1-d)$$

$$P(Y=i)=b(1-a)^{i-1}a=p(1-p)(1-d)^2[1-p(1-d)]^{i-1}, \quad i\geq 1$$

where $Y=i$ means the chain begins with a failure and ends with a failure and in between there are i visits to the states (success). Pr$\{Y=i\}$ can also be directly derived from the original two-state Markov chain with transition matrix P. Since X denotes the total number of units inspected to obtain a non-conforming unit, $X=Y=1$ and the probability function is

$$P(Y=1)=d+p(1-d),$$
$$P(X=i)=p(1-p)(1-d)^2[1-p(1-d)]^{i-2}, \quad i\geq 2 \tag{6.22}$$

The distribution function of X is found as

$$G(x;a,b)=1-(1-p)(1-d)[1-p(1-d)]^{[x]-1} \tag{6.23}$$

We note that when the serial correlation d is zero, the above distribution function reduces to the geometric distribution. The mean and variance are found as

$$\mu=1/p \tag{6.24}$$

and

$$\sigma^2=\left(\frac{1-p}{p^2}\right)\left(\frac{1+d}{1-d}\right) \tag{6.25}$$

It is observed that the mean remains unaffected in the presence of serial correlation, whereas the variance is affected by the quantity $(1+d)/(1-d)$. If a positive serial correlation is present, one will be underestimating the variance of X if a geometric distribution is used and overestimating if d is negative. In other words, the variance of X is larger than the variance of the geometric distribution if d is positive.

6.3.3. An application example

Although our work was originally motivated by a high-quality production environment, our results also hold for any value of p. The following example, although it may not be appropriate from a high quality standpoint, is used for illustrative purpose.

Example 6.4:
Xie and Goh (1993) presented three data sets for read-write error testing of a certain computer hard disk while presenting a modified Poisson distribution to model near zero-defect production processes. The authors observed that these data sets might have some inherent phenomena besides non-conformities caused by random shocks; in particular, the observation might be correlated.

These data sets were examined for serial correlation and we now discuss the result for the third data set in Table 6.5.

Table 6.5. Read-write errors of computer hard disk data (number of non-conformities; read from left to right and down)

4	0	0	0	0	95	80	35	2	0
0	2	0	16	0	0	0	2	0	0
196	0	12	0	0	27	92	117	25	0
0	92	3	0	0	21	188	0	0	0
0	0	362	0	0	0	96	14	0	0
0	0	0	0	0	0	95	0	86	0
0	0	0	0	83	0	221	6	0	0
89	24	0	0	6	26	20	0	0	0
0	0	0	0	0	0	101	0	10	0
0	0	0	0	0	4	68	0	0	0
0	0	0	0	6	2	92	0	20	1
23	0	0	0	0	0	1	101	0	57
98	0	0	0	2	3	312	5	0	0
183	0	0	0	0	1	98	69	1	3
0	0	0	0	1	0	71	0	0	0
16	0	0	0	0	67	119	0	100	0
21	1	0	0	77	86	111	2	0	0
1	98	0	22	95	5	65	1	1	1
0	0	0	0	95	0	7	0	98	0
0	0	1	0	0	0	0	0	1	1
0	0	1	6	92	4	11	0		

Although the process may not really be considered to be a high quality process, it is used here as an illustration of our procedure since it was indicated that it cannot be dealt with using the approach presented in Xie and Goh (1993).

In fact, the data set in Table 6.5 illustrates the problem of clustering of non-conforming items, which is common when data are collected using automated inspection devices. In general, clustering occurs due to clusters in raw materials, accumulation of waste, etc. It is strongly associated with serial dependence and therefore a Markov model is suitable for modeling this type of data.

If the samples are classified as conforming or non-conforming (samples having at least one non-conformity), then the following four possible states would be obtained:

State	Number of events
Non-conforming sample to non-conforming sample	78
Non-conforming sample to conforming sample	43
Conforming sample to non-conforming sample	43
Conforming sample to conforming sample	43

We note that there are 208 entries in Table 6.5. This corresponds to 207 transitions, which is exactly the total tally of the number of events here. Although it was not clear from the data source, we may reasonably presume that the computer concerned has been running for a long time. Therefore, the stationarity of the process in this example may be assumed and thus we can proceed to estimate the parameter p.

It can be found that

$$\hat{p}_{01} = 43/121 = 0.355 \text{ and } \hat{p}_{01} = 43/86 = 0.5$$

and hence

$$\hat{p} = 0.4152 \text{ and } \hat{d} = 1 - \hat{p}_{01} - \hat{p}_{10} = 0.145$$

Furthermore, it can be shown that the asymptotic variance of $\hat{d}$ is obtained as 0.0048 and the 95% confidence interval is as [0.00927,0.2800], conforming that there is a serial dependence in the samples (Lai *et al.*, 2000).

An undesirable feature of this example is the high value of p. Although the value of d is relatively small, it is significantly different from zero and hence should be considered in the model for monitoring of this type of process. This has been investigated in details Lai *et al.* (2000) and interested readers are referred to the original article where some monitoring policies are discussed.

6.4. A model for correlated attribute processes

This section investigates a method of determining the control limits in the case of correlated data. A correlation model based on statistical exchangeability is used for the modeling of dependence. The effect on the control limits due to the correlation and its comparison with the case of independent observations is made. When the correlation is present and the probability is high that an item is nonconforming directly after a nonconforming item is observed, a decision making procedure for process deterioration detection, which is based on counting the cumulative conforming items after one conforming item is also proposed.

6.4.1. A model based on exchangeable Bernoulli variables

Control charting techniques and decision making procedures should be based on appropriate process model. For the case of possibly correlated process, Madsen (1993) considered several existing correlation models and proposed a generalized correlation binomial model, which is based on the exchangeable Bernoulli random variables. The most important aspect is the when the fraction nonconforming is very low, and there is no correlation of the sampled items, this model reduces to geometric distribution, which is the assumption of *CCC* chart. Among the existing correlation models, this model is one of the few that are both realistic and mathematically tractable.

For a high quality process, we define variable X_i as

X_i=0, the i:th item is conforming or

X_i=1, the i:th item is nonconforming.

A sequence $\{X_i\}$ is said to be exchangeable if the following is satisfied

$$P(X_1 = x_1, X_2 = x_2, ..., X_n = x_n) = P(X_1 = x_{a1}, X_2 = x_{a2}, ..., X_n = x_{an})$$

where x_1 equals to 0 or 1 while $\{a_1, a_2, \ldots, a_n\}$ is any permutation of integers $\{1,2,\ldots,n\}$.

Madsen (1993) considered a sequence of exchangeable dependent Bernoulli random variables $\{X_i\}$ which satisfies

$$\prod_k = P\{X_1 = 1, X_2 = 1, ..., X_k = 1\} = \rho p + (1-\rho)p^k \tag{6.26}$$

where p is the fraction nonconforming which is normally less than 0.0001 in high quality processes. The parameter ρ is the correlation coefficient between any pair, ie.,

$$\rho = corr(X_i, X_j),\ i \neq j \tag{6.27}$$

assumed to be greater than or equal to zero.

Usually the parameters are not known. The parameter p is the fractional nonconforming which can be estimated in the traditional way. The method of moment estimator for ρ is

$$\rho = \frac{S^2 - np(1-p)}{n(n-1)p(1-p)} \tag{6.28}$$

where S^2 denotes the sample variance for the observed data defined as

$$S^2 = \frac{\sum_{i=1}^{n}(X_i - \overline{X})^2}{n-1} \tag{6.29}$$

where $\overline{X}$ is the sample mean. In the following, we assume that the model parameters are known.

It is noticed that the two states of outcome in a Bernoulli trial, which are "Success" and "Failure", are mutually dependent and exchangeable. Thus, similarly, we can obtain the following relationship

$$\begin{aligned}\prod{}_k &= P\{X_1 = 0, X_2 = 0, ..., X_k = 0\} \\ &= \rho(1-p) + (1-\rho)(1-p)^k\end{aligned}$$

The parameters p and ρ are the same as above.

If the sequence of the variables $\{X_i\}$ are independent, the correlation coefficient ρ is zero, which is an extreme case. When the sampled items are totally independent, the equation above reduces to the following:

$$\prod{}_k = P\{X_1 = 0, X_2 = 0, ..., X_k = 0\} = (1-p)^k \tag{6.30}$$

As expected, this stands for the geometric distribution, which is the basic assumption of *CCC* chart.

The control scheme in the presence of correlation

Because of the correlation, after one nonconforming item is detected and even when the process is not out of control, the probability for the next item to be nonconforming is larger than the correlation coefficient. Hence, even for a moderate correlation coefficient of 0.1, the probability that there will be a nonconforming item within any positive number of items is greater than any reasonable false alarm probability; *LCL* will normally not exist. This can be seen from the following.

Denote by Y the total number of items to be inspected until the first nonconforming item Y is observed, assuming X_0=1, i.e., the counting starts after a nonconforming item. The probability function of Y is, for i>1,

$$P(Y=i)=P(X_1=0,X_2=0,\ldots,X_{n-1}=0,X_n=1|X_0=1)$$
$$=\frac{P(X_1=0,X_2=0,\ldots,X_{n-1}=0,X_n=1)}{P(X_0=1)}$$
$$=(1-\rho)p(1-p)^{i-1}$$

For i=1, we have

$$P(Y=1)=P(X_1=1\,|\,X_0=1)=\rho+(1-\rho)p \tag{6.31}$$

Hence,

$$P(Y=i)=\begin{cases}\rho+(1-\rho)p, & i=1\\ (1-\rho)p(1-p)^{i-1}, & i\geq 2\end{cases} \tag{6.32}$$

As can be seen from above, $P(Y=1)$ is equal to $\rho+(1-\rho)p$ which is greater than ρ. This implies that if this is greater than the acceptable false alarm probability, the lower control limit will not exist. Notice that Type I error acceptable in practice is usually very small, such as 0.0027 for the traditional Shewhart used in industry. In this case, if there is a small serial correlation, the lower control limit will not exist when the serial correlation ρ is positive. As *LCL* is important for detecting process deterioration, this is not desirable and another monitoring scheme is needed.

One way to resolve this problem is to look at the number of conforming items after the first conforming item. That is, if the first item is nonconforming after a nonconforming item, it might be caused by the correlation; this has to be checked. On the other hand, if we have one conforming item, then the number of consecutive conforming items before another nonconforming item should be large.

6.4.2. Determination of the lower control limit

Consider a high quality manufacturing process that produces individual items, each of which is inspected in the order of production. The fraction nonconforming is usually very small. Using the model described previously, the probability that a nonconforming item is not found in n inspected items, given the condition that the first item is conforming, is stated as the following:

$$\begin{aligned} &P(\text{no nonconforming in } n \text{ items} \mid \text{the first item is conforming}) \\ &\quad = P\{X_1 = 0, X_2 = 0, \ldots, X_{n-1} = 0, X_n = 0 \mid X_1 = 0\} \\ &\quad = [\rho(1-p) + (1-\rho)(1-p)^n]/(1-p) = \rho + (1-p)(1-p)^{n-1} \end{aligned}$$

In a high quality process, we usually pay more attention to process deterioration. Thus, we usually concentrate on the *LCL* of *CCC* chart instead of *UCL*. Now, we introduce the confidence level s so that (1-s) is the probability that the process is out of control (in fact the process has deteriorated) when a point is plotted below *LCL*. It is also common to use a value other than 0.00135 which is associated with traditional Shewhart chart for this probability in high quality processes.

Hence, we have the relation between the certainty level s, the fraction nonconforming p, the correlation coefficient ρ, and the number of conforming items inspected n, given as

$$s = \rho + (1-\rho)(1-p)^{n-1} \tag{6.33}$$

We can now determine the number of conforming items inspected before a nonconforming one is allowed, which is actually (*LCL*+1) in the presence of correlation. Denote by LCL_i and LCL_c the *LCL* without correlation and with correlation respectively. LCL_c can be calculated as

$$LCL_c(\rho) = \frac{\ln\left(\dfrac{s-\rho}{1-\rho}\right)}{\ln(1-p)} \tag{6.34}$$

A comparison of LCL_c and LCL_i

The lower control limit, *LCL*, when the inspected items are independent of each other is given as

$$LCL_i = \frac{\ln s}{\ln(1-p)} \tag{6.35}$$

It is obvious that when the correlation coefficient ρ is equal to zero, which implies that the items are independent, we have the following equality:

$$LCL_i = LCL_c(0) = \frac{\ln s}{\ln(1-p)} \tag{6.36}$$

It is of interest to see how much the LCL_c and LCL_i differ in practice, as this is an indication of possible errors when the independence assumption is used. Hence, the relationship between LCL_c and LCL_i is studied here and the difference between the two is given as

$$LCL_c - LCL_i = \frac{\ln(\frac{s-\rho}{1-\rho})}{\ln(1-p)} - \frac{\ln s}{\ln(1-p)} = \frac{\ln\frac{s-\rho}{s(1-\rho)}}{\ln(1-\rho)} \tag{6.37}$$

It can be shown that this is positive if ρ is positive and vice versa. Hence the conventional *CCC* chart will not be sensitive enough when the correlation is positive, while it will give fewer alarms when the correlation is negative.

We can also consider the relative difference between LCL_c and LCL_i. This is given as

$$\frac{LCL_c}{LCL_i} = \frac{\ln\frac{s-\rho}{1-\rho}}{\ln(1-p)} / \frac{\ln s}{\ln(1-p)} = \ln\frac{s-\rho}{1-\rho} / \ln s \tag{6.38}$$

Note that this ratio is independent of p. A conclusion that can be drawn is that if the *LCL* for the independent case is used when there is a positive correlation, it could be much smaller than the *LCL* that should have been used.

6.5. Some issues for further research

Although geometric distribution could be directly used for the count of defects or nonconformities (e.g., Kaminsky *et al.*, 1992 and Glushkovsky, 1994), our focus is on the use of this distribution in the modeling of the number of items inspected before a nonconforming item is observed. This is different from the traditional Poisson model used for the count of defects or nonconformities of products. For this type of count data, when the Poisson distribution is not appropriate, extension of the Poisson model could be considered.

The model of ZD process with random shock presented here is also known as zero-inflated Poisson model (Xie *et al.*, 2001c). This model can be used for count data when there is a larger dispersion compared with the Poisson distribution. Another flexible model is the generalized Poisson distribution (Consul, 1989). It could be of interest to investigate the application of this and other extended Poisson distributions for quality related data modeling and analysis.

The data correlation is usually a difficult problem. Although some models and analysis are presented here, we have avoided presenting the detailed monitoring procedure. In fact, for serially correlated data, some kind of moving sum approach could be suitable. Some discussions can be found in Lai *et al.* (2000).

A limitation of the original *CCC* chart is that the decision is based on a single count data. There are many possible ways to incorporate earlier information. The *CCC-r* chart is a simple method. The conditional procedure is another possibility. Methods such as *EWMA* and *CUSUM*

CHAPTER 7

CUSUM and *EWMA* Procedures

A problem with the conventional Shewhart charts is that the decisions are usually based on single observations. Such decision implies a high probability of false alarms. Also, small shifts in the process characteristics, which are of increasing importance from the point of view of efforts for continuous quality improvement, cannot be detected easily. Cumulative sum (*CUSUM*) and exponentially weighted moving average (*EWMA*) charts are widely recommended. Although they are developed preliminary for variable data, several authors have investigated their properties for attribute data for which the *CCC* chart is related to.

The *CUSUM* chart is based on the charting of cumulative sum of previous observations which allows us to use all the information about the process to make more accurate decisions. Similar to the *CUSUM*, the *EWMA* is also an effective method for quick shift identifications for small process shifts. Thus, in the following sections these two are discussed in detail with application examples. It should be pointed out that the *CUSUM* is not related to the *CCC* chart and the *CUSUM* can be used with cumulative count as it is used for other count or variable data. We will first describe the traditional approach and then their uses on *CCC* data as an application. Some recent research results are also presented.

7.1. General *CUSUM* procedure

The *CUSUM* technique was first developed in 1950s as an alternative to the Shewhart chart (Page, 1954, 1961). It has been used in a number of companies and there are many published research articles on it, see e.g., Hawkins (1981), Gan (1991), Woodall and Adams (1993), Prabhu *et al.* (1997), Atienza *et al.* (2000), Luceno and Puig-Pey (2000) and Bourke (2001).

The use of the traditional *CUSUM* chart based on the use of V-mask in decision making is discussed and a decision interval scheme for *CUSUM* is presented. There are many other charting procedures available. The V-mask procedure is relatively complicated, as an actual V-mask has to be used and moved around. Although it is not recommended, for the sake of completeness we summarize the procedures here. Details can be found in many standard texts on statistical process control, see e.g., Montgomery (2001).

7.1.1. *CUSUM* with V-Mask

When calculating the *CUSUM*, a reference value T is needed. This can either be the average value of the previous measurement or the target value of the process, usually applicable for variable data. The cumulative sum Y_n is calculated as the sum of the total deviation from this target value by the measurement values X_i, i=1,2,...,n, i.e.,

$$Y_n = \sum_{i=1}^{n}(X_i - T) \tag{7.1}$$

As it can be seen, the effect of systematic change in the process characteristic is accumulated in Y_n. Hence, the decision based on the Y_n is actually based on all previous measurement values X_i, i=1,2,...,n.

For decision making using the traditional V-mask, or the truncated V-mask, there are two parameters, h and f determining the shape of the V-mask. Usually they are set to be

$$h=5 \quad \text{and} \quad f=1/2, \tag{7.2}$$

respectively, although other values can be used depending on the average run length that can be modified in practice. This is similar to the case of 3-sigma limits for which the value three is commonly used. A charting procedure is shown in Figure 7.1.

The *CUSUM* procedure is good when small shifts of process mean have to be detected and measurement is expensive. Compared with the conventional Shewhart charts *CUSUM* procedure has shown to be effective in detecting small process shifts and has less number of false alarms when the process is on target. It is less effective than the Shewhart chart when only large process shifts are of importance and it is more complex to use, leading to more human errors.

Example 7.1:

To illustrate the use of *CUSUM* chart and its comparison with the traditional Shewhart chart, an example is shown here where nonconformity data is summarized in Table 7.1. The traditional c-chart with three-sigma limits is shown in Figure 7.2.

Table 7.1. A set of nonconformity data

4	4	1	9	2	5	4	8	4	3
3	6	4	4	9	5	2	5	9	3
5	1	4	3	6	4	8	6	4	6
7	9	9	8	1	11	8	7	12	5
11	13	5	11	3	9	6	4	8	12

Table 7.2. The *CUSUM* for the nonconformity data in Table 7.1

-1	-2	-6	-2	-5	-5	-6	-3	-4	-6
-8	-7	-8	-9	-5	-5	-8	-8	-4	-6
-6	-10	-11	-13	-12	-13	-10	-9	-10	-9
-7	-3	1	4	9	15	18	20	27	27
33	41	41	47	45	49	50	49	52	59

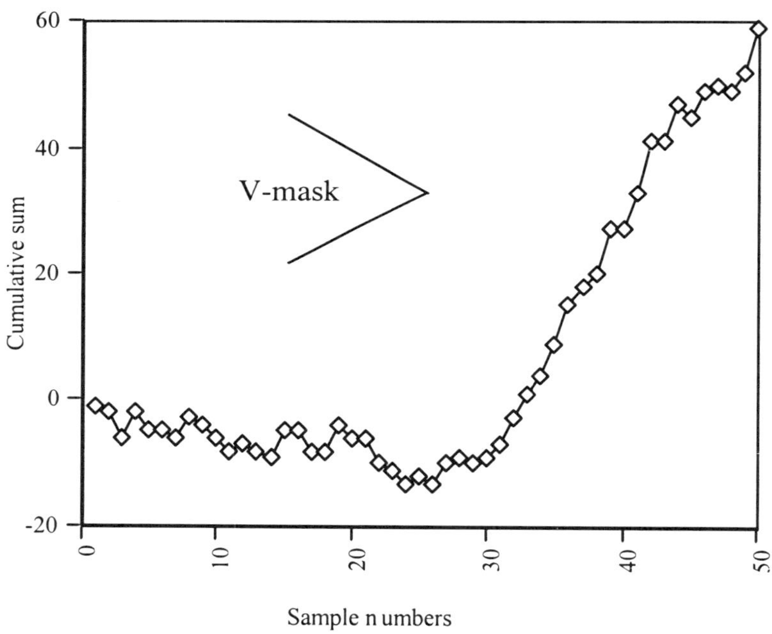

Figure 7.3. The *CUSUM* chart for the nonconformity data in Table 7.2. It is clear that the process has changed at about the sample number 30.

In fact, the values presented in Table 7.2 are simulated with first 30 data points as a Poisson sample with mean equal to 5, while the last 20 values are simulated Poisson sample with mean equal to 10, twice as large as the former. The traditional Shewhart chart cannot detect this change while the *CUSUM* chart is able to do it.

□

7.1.2. *CUSUM* with decision interval

The standard procedure for *CUSUM* charting and decision making is related to the use of a V-mask which is not practical. It has a problem when the procedure is computerized although it is easy to use the V-mask for illustrative purposes. Because there is no similarity to the conventional Shewhart charts, which are well-known and wide-spread in industry, decision making is difficult to be explained. On the other hand, a decision interval scheme is available for *CUSUM* procedure, which is statistically similar to the conventional approach of *CUSUM* charting, but with a layout and control limit similar to those in traditional Shewhart charts.

The procedure based on the decision intervals only exists for detecting changes in one direction. The amount of change in each direction is compared with a predetermined amount that is the decision interval h. The process is judged out of control whenever the value exceed h. A two-sided method, which is based on the plotting on two charts, can be used.

The *CUSUM* value C_1 can be calculated similar to (7.1), but by subtracting

$$K_u = T + f\sigma \tag{7.3}$$

instead. Here T is the reference value as before and f defines the slope of the V-mask, which is traditionally taken as f=1/2. The *CUSUM* value is kept to be positive and whenever a new series starts when it falls below zero.

Similarly, we define

$$K_l = T - f\sigma \tag{7.4}$$

and calculate the *CUSUM*, C_2, by subtracting K_l and keep it to be negative. Action is signaled if the size of either *CUSUM* series reaches or exceeds $H=h\sigma$.

Example 7.2:
We illustrate the use of decision interval scheme by applying it to the data set in Table 7.1. The plot is shown in Figure 7.4 and it is clear that the out of control is signaled at the sample number 35, i.e. 5 samples after the change of the process mean.

□

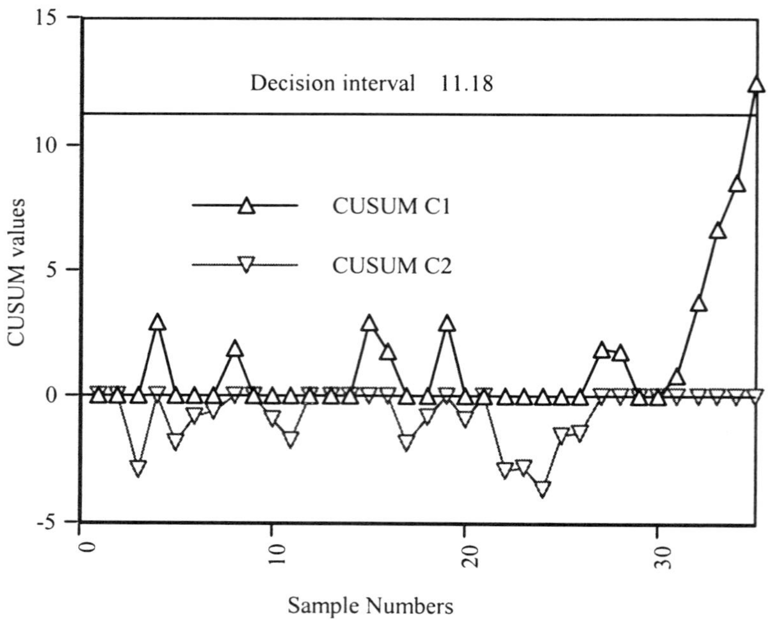

Figure 7.4. The *CUSUM* based on the decision interval scheme for the data in Table 7.1.

7.1.3. *CUSUM* chart performance

The performance of *CUSUM* and related techniques is an important area of research. The derivation of the *ARL* of a *CUSUM* procedure is relatively difficult compared to Shewhart-type control charts. Many authors have contributed on this topic since its introduction. For continuous observations,

Page (1954) provided simple integral equation and recursive equations for the *ARL* function and probability function of run length of a *CUSUM* chart respectively. Johnson (1961) used Wald's approximation to determine the *ARL* utilizing the probabilities of Type I and Type II errors of hypothesis testing for the change of process mean. von Dobben de Bruyn (1968) tabulated simulation results of *ARL*'s for *CUSUM* procedures with different decision intervals and reference values. Goel and Wu (1971) created a nomogram for *ARL*, decision intervals, and reference values by solving systems of linear equations numerically.

Brook and Evans (1972) used the Markov chain approach to approximate the moments and percentages points of run length distribution for integer-valued *CUSUM* procedures. Reynolds (1975) derived an expression for *ARL*s by Brownian motion approximation. For an application of Markov chain approach, Section 7.2 discusses the use of it in the study of *CUSUM* applied to *CCC* data.

It can be noted that *CUSUM* procedure has mostly widely been used for variable data, i.e. observation that can take any values. Many references are available and recent research focuses on the multivariate and correlated data. Some research on CUSUM procedure for count related data has been presented, among others, in Gardiner (1987), Lucas (1985, 1989), Gan (1993), Radaelli (1994), White and Keats (1996), Reynolds and Stoumbos and Burke (2001).

Lucas (1985) considered a scheme that gives an alarm signal if there are two or more defects within an interval of m samples. This control scheme is also shown to be equivalent to a *CUSUM* chart with reference value being $K = 1/m$ and the threshold value being $H = 1$. The chart for detecting an increase in the defect rate of a process plots the cumulative sum S_i defined as

$$S_i = \max[0, y_i - K + S_{i-1}]$$

where y_i is the inspection result of the i:th sample. If the ith sample contains no nonconforming items, y_i takes the value of 0 and it takes the value of 1 otherwise. An out-of-control signal is given when $S_i > H$. Using the Markov chain approach, the *ARL* of this *CUSUM* chart can be shown to be

$$ARL = \frac{1 - q + p(1 - q^{m-1})}{(1-q)[p(1-q^{m-1}) + u]}$$

where q is the probability of no nonconforming items occurring, p is the probability of the occurrence of a single nonconforming item while u is the probability of more than one nonconforming item.

It should be mentioned that when the process quality level is very high, this procedures would signal at every sample that contains a nonconforming item because they are essentially a *CUSUM* application to the conventional *c*- or *np*-chart. Moreover, it is difficult to detect the decrease in the defect rate. Some further investigation and research is presented in Lucas (1989). On the other hand, the basic *CCC* chart is more appropriate and easier to use. Nevertheless, the use of *CUSUM* idea is useful when applied to *CCC* data itself and will be discussed in more details in Section 7.2.

7.2. *CUSUM* with *CCC* data

The *CUSUM* procedure can be used for geometric distributed data. Recall that when *CCC* chart is used, the plotted point is the number of items checked till a nonconforming one is observed. A general approach using *CUSUM* on *CCC* data is described here.

7.2.1. *CUSUM-CCC* chart

A *CCC* chart or Run-Length control chart (Bourke, 1991) is based on plotting the run lengths between successive nonconforming items, but it can also be generally defined as the measurement before some limit is exceeded. A procedure called *CUSUM-CCC*, which applies the *CUSUM* procedure to the run lengths between successive nonconforming items, is presented here.

Bourke (1991) proposes the Run-Length control chart, based on the plotting of the number (run length) of conforming items between successive nonconforming items, to replace the conventional *p*-chart for situations where continuous production and 100% inspection in the order of products is in progress. The Run-Length control chart, similar to any other Shewhart charts, leads to decisions based only on the latest plotted point and ignores any information about the process contained in the previously plotted points.

Let Y_i denote the *CCC* count between the *i*:th and the (*i*+1):th nonconforming item. If the items are assumed to be independent of each other, Y_i follows independent identical geometric distributions with parameter p. As in any other *CUSUM* procedure, two parameters, which are the reference value K and the decision interval H, are introduced in the *CUSUM-CCC* procedure.

7.2.2. *CUSUM-CCC* chart performance

The study of chart performance is an important issue. Although the average run length (*ARL*) is a suitable measure to evaluate the performance, the number of points plotted on the *CUSUM-CCC* chart does not directly correspond to the number of items inspected. Another term, average number inspected (*ANI*), is used in Bourke (1991) to refer to the average number of items inspected until an alarm signal is produced. The *ANI* concept, which is similar to *AIRL* denoting the run length counted by the number of items inspected before an alarm is raised, will be used here.

The Markov chain approach, which is suitable for discrete observations and provides exact *ARL* values, can be used to derive the *ANI* for the *CUSUM-CCC* procedure. The upper-sided (lower-sided) *CUSUM-CCC* procedure can be regarded essentially as a Markov chain with S_i (T_i) being the state space while $\{S_t > H \text{ or } T_t < -H\}$ being the absorbing state.

(a) The case of process improvement

When process improvement detection is of interest, the upper-sided *CUSUM-CCC* procedure is to be used. Let *L* be the index at which the cumulative sum S_L exceeds the decision interval *H*, while all the sums $S_i (i < L)$ are smaller than *H*. Assume that the reference value *K* and the decision interval *H* are written as $K = a/b$ and $H = c/b$, where *a*, *b* and *c* are positive integers such that *K* is a rational number in the reduced form. This is because the exact *ANI* results can be obtained using a rational *K* and rational numbers are rich enough to provide the practical range of *K* for implementation.

With these notations, the possible values of process state S_i are given by $\{0,\ 1/b,\ 2/b,\ \ldots,\ c/b\}$ when the process is in control, and they are the transient states of the Markov chain. When $S_L > c/b$, the process is out of control and then S_L is absorbing state of the process. The transient states are labeled here as {1, 2, ..., *c*+1}, which correspond to $\{S_i = i/b, i = 0,1,2,...,c\}$ respectively. The absorbing state S_L is labeled as state *c*+2.

The one-step transition probability matrix of the Markov chain is given by:

$$\begin{bmatrix} R & \tilde{P} \\ \tilde{O}^T & 1 \end{bmatrix} = \begin{bmatrix} p_{1,1} & p_{1,2} & \cdot & \cdot & \cdot & p_{1,c+1} & p_{1,c+2} \\ p_{2,1} & p_{2,2} & \cdot & \cdot & \cdot & p_{2,c+1} & p_{2,c+2} \\ \cdot & \cdot & & & & \cdot & \cdot \\ \cdot & \cdot & & & & \cdot & \cdot \\ \cdot & \cdot & & & & \cdot & \cdot \\ p_{c+1,1} & p_{c+1,2} & \cdot & \cdot & \cdot & p_{c+1,c+1} & p_{c+1,c+2} \\ 0 & 0 & \cdot & \cdot & \cdot & 0 & 1 \end{bmatrix}$$

where $p_{i,j}$ denotes the probability of the process moving from state *i* to state *j* in exactly one step. They can be determined using the run-length probability distribution.

Denote vector $(\mathbf{I} - R)^{-1}\mathbf{1}$ by

$$(\mathbf{I} - R)^{-1}\mathbf{1} = (\mu_1, \mu_2, ..., \mu_{c+1})^T$$

in which **I** is the $(c+1)\times(c+1)$ identity matrix while **1** is the unit vector $[1,1,\ldots,1]_{1\times(c+1)}$. The expected value of index *L* can be obtained (Lu, 1999), as

$$E(L) = \mu_1 \tag{7.8}$$

Therefore, the *ANI* for the upper-sided *CUSUM-RL* procedure, ANI_u, is derived as

$$ANI_u = E(L)\cdot E(Y_i) = \mu_1 / p \tag{7.9}$$

(b) The case of process deterioration

Similarly, when process deterioration detection is of interest, the lower-sided *CUSUM-CCC* procedure is to be applied. Let L^* be the index at which the cumulative sum T_{L^*} exceeds the decision interval (-*H*), while all the sums $T_i (i < L^*)$ are larger than (-*H*). In this case, the process transient states are $\{T_i = -i/b, i = 0, 1, 2, ..., c\}$. They are labeled as state {1, 2, 3, …, *c*+1} and the absorbing state is T_{L^*}, which is also labeled as state *c*+2. The one step probability transition matrix is given as

$$\begin{bmatrix} R^* & \tilde{P}^* \\ \tilde{O}^T & 1 \end{bmatrix} = \begin{bmatrix} p^*_{1,1} & p^*_{1,2} & \cdot & \cdot & \cdot & p^*_{1,c+1} & p^*_{1,c+2} \\ p^*_{2,1} & p^*_{2,2} & \cdot & \cdot & \cdot & p^*_{2,c+1} & p^*_{2,c+2} \\ \cdot & \cdot & & & & \cdot & \cdot \\ \cdot & \cdot & & & & \cdot & \cdot \\ \cdot & \cdot & & & & \cdot & \cdot \\ p^*_{c+1,1} & p^*_{c+1,2} & \cdot & \cdot & \cdot & p^*_{c+1,c+1} & p^*_{c+1,c+2} \\ 0 & 0 & \cdot & \cdot & \cdot & 0 & 1 \end{bmatrix}$$

where $p^*_{i,j}$ denotes the probability of the process moving from state *i* to state *j* in one step.

Similar to the case of process improvement, the *ANI* for the lower-sided *CUSUM-CCC* procedure, ANI_l, can be obtained as

$$ANI_l = E(L^*) \cdot E(Y_i) = \mu^*_1 / p \tag{7.10}$$

where $E(L^*) = \mu^*_1$ is the first element of vector $(\mathbf{I} - R^*)^{-1}\mathbf{1}$.

7.2.3. Determination of chart parameters

A *CUSUM* control scheme is equivalent to a sequence of Wald sequential probability ratio test (*SPRT*). This allows the optimal design of *CUSUM* procedures to be developed (Moustakides, 1986). For a process being monitored, denote p_a the acceptable process fraction nonconforming and p_d the fraction nonconforming that the *CUSUM-CCC* procedure is to detect promptly.

Let the null hypothesis be $H_0 : p = p_a$ and the alternative hypothesis be $H_1 : p = p_d$. Hence, for the geometric counts Y_i, the *SPRT* employs the likelihood ratio sequentially as

$$\lambda_m = \frac{\prod_{i=1}^{m}(1-p_a)^{Y_{i-1}} p_a}{\prod_{i=1}^{m}(1-p_d)^{Y_{i-1}} p_d}, \quad m = 1,2,... \tag{7.11}$$

An equivalent test involving the natural logarithm of λ_m is often used and $\ln(\lambda_m)$ can be expressed as

$$\ln(\lambda_m) = \sum_{i=1}^{m-1}(Y_i - K_{SPRT}) + Y_m - K_{SPRT} \tag{7.12}$$

where the reference value of the *SPRT*, K_{SPRT}, is given by

$$K_{SPRT} = \frac{\ln\{[p_a(1-p_d)]/[p_d(1-p_a)]\}}{\ln[(1-p_d)/(1-p_a)]} \tag{7.13}$$

This test continues as long as $A_0 < \ln(\lambda_m) < A_1$ for some fixed values A_0 and A_1. When $p_d > p_a$, the test accepts H_0 if $\ln(\lambda_m) > A_1$ and rejects H_0 if $\ln(\lambda_m) < A_1$; while when $p_d < p_a$, it accepts H_0 if $\ln(\lambda_m) < A_1$ and rejects H_0 if $\ln(\lambda_m) > A_1$.

Thus, the optimal reference value of the corresponding *CUSUM-CCC* procedure, *K*, is

$$K = \frac{\ln\{[p_a(1-p_d)]/[p_d(1-p_a)]\}}{\ln[(1-p_d)/(1-p_a)]} \tag{7.14}$$

and the standardized reference value *k* is available with

$$k = \frac{p_a}{\sqrt{1-p_a}} \cdot [1 - \frac{\ln p_a - \ln p_d}{\ln(1-p_a) - \ln(1-p_d)}] \tag{7.15}$$

Some numerical values of k and K are tabulated in Table 7.4.

Table 7.4. Some numerical values of parameters K and k for the *CUSUM-CCC* procedure

p_d / p_a	p_a							
	0.0005		0.001		0.005		0.01	
	K	k	K	k	K	k	K	k
1.1	1906.2	0.9533	953.1	0.9536	190.6	0.9555	95.3	0.9579
1.2	1823.2	0.9118	911.6	0.9121	182.3	0.9139	91.2	0.9162
1.3	1749.1	0.8748	874.5	0.8750	174.9	0.8767	87.4	0.8789
1.4	1682.4	0.8414	841.2	0.8416	168.2	0.8432	84.1	0.8453
1.5	1621.8	0.8111	810.9	0.8113	162.2	0.8129	81.1	0.8149
2.0	1386.3	0.6933	693.1	0.6935	138.6	0.6947	69.3	0.6962
5.0	804.5	0.4024	402.2	0.4024	80.3	0.4023	40.0	0.4022
10.0	511.3	0.2557	255.4	0.2556	50.8	0.2544	25.2	0.2529
15.0	386.3	0.1932	192.9	0.1930	38.1	0.1911	18.8	0.1886
20.0	314.7	0.1574	157.0	0.1571	30.9	0.1547	15.1	0.1513
40.0	188.2	0.0941	93.6	0.0937	17.9	0.0898	8.4	0.0841
80.0	109.7	0.0549	54.2	0.0542	9.7	0.0484	3.7	0.0376
p_d / p_a	p_a							
	0.0005		80		0.005		0.01	
	K	k	K	k	K	k	K	k
0.90	2107.2	1.0539	1053.6	1.0541	210.7	1.0562	105.4	1.0589
0.80	2231.4	1.1160	1115.7	1.1163	223.1	1.1185	111.6	1.1213
0.70	2377.8	1.1892	1188.9	1.1895	237.8	1.1918	118.9	1.1948
0.60	2554.1	1.2774	1277.0	1.2777	255.4	1.2802	127.7	1.2833
0.50	2772.5	1.3866	1386.3	1.3869	277.2	1.3896	138.6	1.3929
0.10	5116.4	2.5589	2558.0	2.5593	511.3	2.5628	255.4	2.5672
0.05	6306.1	3.1539	3152.7	3.1543	630.0	3.1580	314.7	3.1627
0.02	7982.7	3.9923	3990.8	3.9928	797.3	3.9967	398.1	4.0015
0.01	9302.0	4.6522	4650.3	4.6527	929.0	4.6566	463.8	4.6615

In order to design a one-sided *CUSUM-CCC* procedure for monitoring a statistic, we need to know all design parameters: the in-control average run length $ARL(0)$, the reference value K and the decision interval H. The most common way is to find an adequate decision interval H based on the other parameters. The *ARL* at a target value is a measure of how frequently an alarm signal is produced on average in an acceptably running process. The idea of giving this $ARL(0)$ is the same as giving Type I error α in hypothesis testing and their relationship is

$$ARL(0) = 1/\alpha \tag{7.16}$$

However, the average number inspected is used instead of the conventional average run length in the *CUSUM-CCC* procedure because of the reasons mentioned in previous sections. As the *CCC* count between two successive nonconforming items is independent on the number of plotted points on the *CUSUM-CCC* procedure until an alarm signal appears, it is not difficult to understand that the in-control average number inspected for an one-sided *CUSUM-CCC* procedure is

$$ANI_u(0) = ANI_l(0) = E(Y_i) \cdot ARL(0) = 1/(p_a \alpha) \tag{7.17}$$

Theoretically, we would like the process to run for as long as possible when it is at an acceptable level, and to interfere with the process as soon as possible when it is shifting away from the target level. In other words, the *ANI* is expected to be appropriately large when process fraction nonconforming p is at the acceptable value and appropriately small when p is at the value that is to be detected quickly by the *CUSUM-CCC* procedure.

Furthermore, we also have

$$ANI_u(0) = \mu_1 (1 + \frac{1}{p_a}) \tag{7.18}$$

and

$$ANI_l(0) = \mu^*_1 (1 + \frac{1}{p_a}) \tag{7.19}$$

(1) Determine the acceptable fraction nonconforming p_a and the out-of control fraction nonconforming p_d for which a quick detection is desirable;

(2) Choose the in-control average number inspected or type I error;

(3) Determine the optimal reference value k and K;

(4) Determine the decision interval h and H such that the *CUSUM-CCC* control procedure has an in-control *ANI* or Type I error which are specified in Step 2;

(5) Compute the cumulative sums S_i and T_i, whenever a nonconforming item is detected;

(6) Make the decision that whether the process has deteriorated or improved using the decision rule introduced before.

The *CUSUM* procedure applied to *CCC* data has been compared with the original *CCC* procedure in Xie *et al.* (1998). In general, as expected, the *CUSUM* applied to *CCC* improves the performance of small shift.

7.3. Basics of *EWMA* charts

While the decision making based on traditional Shewhart charts relies on single observation, the *CUSUM* procedure assumes that all previous data are equally informative about the current state of the process. Both are extreme cases. The use of weighted moving average control scheme should then be a better alternative. One popular moving average chart is the so-called exponentially weighted moving average chart, or *EWMA* chart. It was first introduced in Roberts (1959) who showed that it is useful for detecting small shifts in the mean of a process. This type of control charts have, however attracted more interest in recent years, see e.g., Lucas and Saccucci (1990), McGregor and Harris (1993),

Wasserman (1995), Van Brackle and Reynolds (1997), Zhang (1998) and Chen *et al.* (2001). Gan (1990) and Borror (1998) are some of the papers that deal with *EWMA* applied to traditional binomial or Poisson count data.

The *EWMA* control scheme is based on the weighted information of previous measurement and it has a very simple form. An *EWMA* has also been referred to as a geometric moving average. We can note that the form for Z_i is equivalent to a moving average of current and past observations with the weights of the past observations decrease exponentially as in a geometric series, i.e.,

$$Z_i = \lambda \sum_{j=0}^{i-1} (1-\lambda)^j X_{i-j} + (1-\lambda)^i Z_0 \tag{7.21}$$

Usually the 3-sigma type control limits are used also for *EWMA* chart. It is well-known that if Y_i's are independent identically distributed with a common standard deviation, σ, then the variance for *EWMA* Z_i is given by

$$\sigma_{Z_i}^2 = \frac{\lambda(1-(1-\lambda))^{2i}}{2-\lambda}\sigma^2 \tag{7.22}$$

Note that this depends on the sample number i. However, it converges to an asymptotic value given by

$$\sigma_Z^2 = \frac{\lambda}{2-\lambda}\sigma^2 \tag{7.23}$$

Traditionally this is used in determining the control limits using the following formula

$$\begin{aligned} UCL &= E(X) + 3\sigma_Z \\ CL &= E(X) \\ LCL &= E(X) - 3\sigma_Z \end{aligned} \tag{7.24}$$

where $E[X]$ is the target value of the process characteristic to be monitored. Note that instead of 3, other values could be used. The selection of this for *EWMA* can be found in some standard texts such as Montgomery (2001). The charting procedure can be summarized in Figure 7.6.

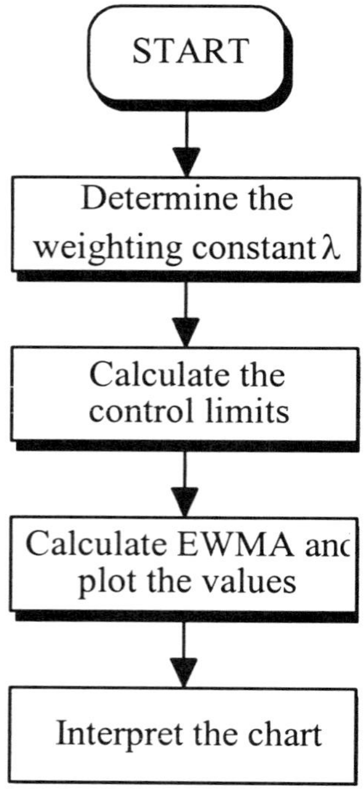

Figure 7.6. Steps in using *EWMA* chart.

Note that the parameter λ determines the weight on the historical data; this is the major difference between the traditional Shewhart charts, the *CUSUM* chart and the *EWMA* chart. While for the Shewhart chart, signals for process out of control is based on the last data point and none of the previous data points are made use of, both the *CUSUM* chart and the *EWMA* chart depend upon a sum to determine if the process is out of control. Equal weight is given to all past data points in the *CUSUM* chart, but for the *EWMA* chart, less and less weight is given to older data.

The weight heavily depends on the value of λ and if λ is close to 1, the *EWMA* tends to approach the traditional Shewhart chart. On the other hand, if λ is close to 0, then the *EWMA* has a long "memory" and it approaches a *CUSUM* chart with respect to its ability to detect a process change. The selection of λ is a practical problem and different schemes have been proposed for this purpose. One of the straightforward methods is through the minimization of the sum of squared errors when the model is fitted to past data.

EWMA is a time series chart as well as a forecasting based monitoring technique. It is particularly suited for correlated data. Detailed discussions can be found in Alwan and Roberts (1988), Alwan (1992), Kramer and Schmid (1997), Jiang *et al.* (2001), among others. See also Box *et al.* (1994) for a general treatment of time series forecasting technique.

The design of *EWMA* can be based on the average run length which has to be obtained by simulation or approximation methods. Such tables have been developed by some authors, see e.g., Lucas and Saccucci (1990).

Example 7.4:
The application of *EWMA* on *CCC* data is straightforward. For the data given in Table 7.3, the corresponding *EWMA* chart with λ=0.2 is as shown in Figure 7.7. Note that in this *EWMA* chart, the exact control limits, which are suitable for small sample numbers, are used.

□

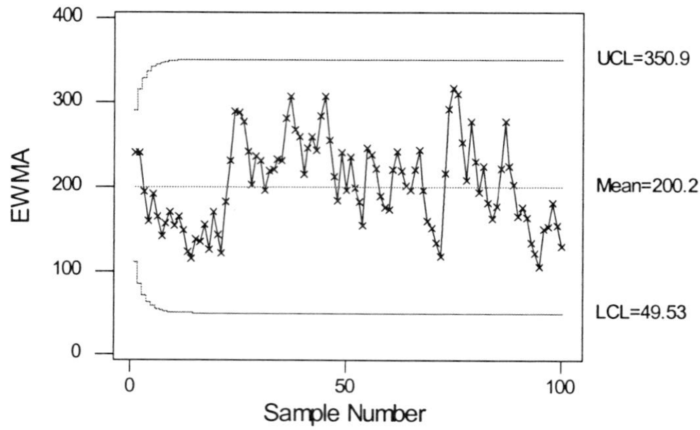

Figure 7.7. An EWMA chart for the data set in Table 7.3.

7.4. *EWMA* with *CCC* Data

EWMA idea can be applied to *CCC* data to improve the ability to detect small shifts. However, because the distribution of the *CCC* data is highly skewed, it would be useful to transform the data to one close to normal distribution. There are several possibilities transformations that can be used with *EWMA* procedure. One is the geometric *EWMA-Q* chart, which is based on a transformation in Quesenberry (1995). Another one is the geometric *EWMA*-$X^{1/4}$ chart, which is based on the double square root transformation of *CCC* data. The latter has the advantage that the distribution is not assumed to be known or estimated. We will focus on this in this section.

7.4.1. *EWMA* with transformation

The square root transformation can be used for positively skewed data, but in this case the geometric distribution is very skewed, so that a double square root transformation is suggested. The chart is called *EWMA*-$X^{1/4}$ chart here.

The *EWMA* for double square root statistic D_i can be given by:

$$z_i = \theta D_i + (1-\theta) z_{i-1} \tag{7.25}$$

where

$$D_i = (X_i)^{0.25} \tag{7.26}$$

The centerline and control limits for the geometric *EWMA*-$X^{1/4}$ control chart can be computed as follows:

$$UCL = u_0 + 3\sigma_D \sqrt{\frac{\theta}{2-\theta}} \tag{7.27}$$

and

$$LCL = u_0 - 3\sigma_D \sqrt{\frac{\theta}{2-\theta}} \tag{7.28}$$

Here u_0 is the target value of the double square root transformed *CCC* data.

Example 7.5:

To illustrate the usage of the *EWMA* for double square root transformed *CCC* data, a set of data from Quesenberry (1995) is used. The first ten values are simulated from geometric distribution with $p = 0.0001$ and the next ten with $p = 0.001$. The chart is given in Figure 7.8. It clearly illustrates a typical pattern that results from an increase in p. Compared to the geometric *EWMA-Q* chart, which is based on the inverse normal transformed *CCC* data, the geometric *EWMA*-double square root chart is simpler to implement and interpret. Their performances in detecting the shift are almost the same. Therefore, from a practical point of view, the latter is more appropriate for monitoring high quality processes.

□

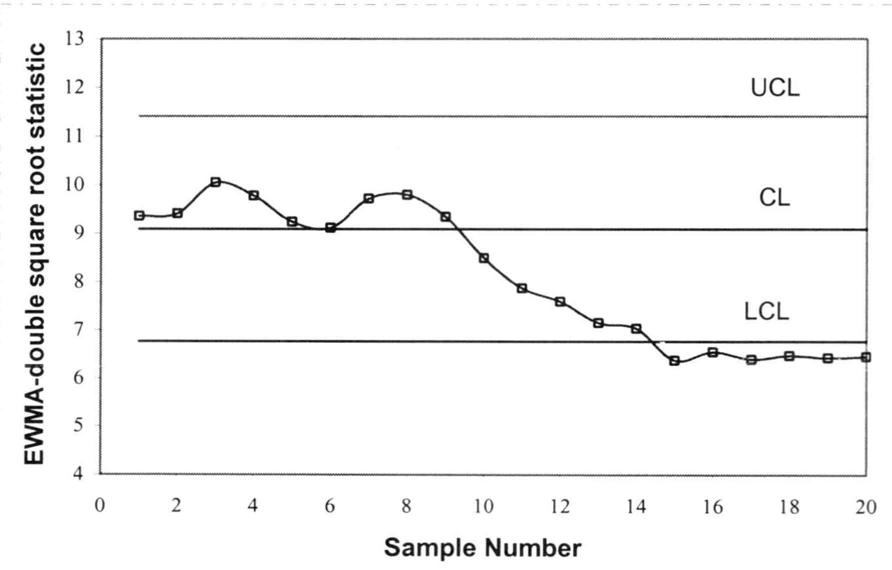

Figure 7.8. A geometric *EWMA*-double square root chart.

7.4.2. Some statistical properties

In this section we compute the average run length for the $EWMA\text{-}X^{1/4}$ chart. The ARL is defined here as the average number of individual items that must be observed before an out-of-control signal is indicated. A good control scheme should have a high sensitivity, i.e., a low ARL when an undesirable factor is affecting the process and a high ARL when the process is stable.

The ARL is easily calculated for independent process characteristics. However, since the *EWMA* statistic is positively autocorrelated, the computations of an ARL is more complex. Many works address the ARL performance of the *EWMA*. For instance, Roberts (1959) studied the problem using simulation techniques in the early *EWMA*. Robinson and Ho (1978) developed a numerical technique using an Edgeworth series expansion approximation. Lucas and Saccucci (1990) modeled the *EWMA* control schemes as a Markov chain, which allows a simpler and more accurate determination of an ARL.

Crowder (1987) evaluated the ARL properties of *EWMA* by formulating and solving a system of integral equations. The integral equation is:

$$L(\mu) = ARL = 1 + \frac{1}{\theta} \int_{LCL}^{UCL} \frac{L(x) f(x - (1-\theta)\mu)}{\theta} dx \tag{7.29}$$

where $f(x)$ is the $N(\mu, \sigma^2)$ probability density function and the *EWMA* starts at μ.

A numerical procedure is employed to approximate the above integral equation and the corresponding computer program is provided in Crowder (1987). The computer program can be used to calculate the ARL of the *EWMA*-$X^{1/4}$ for known θ, L, and δ. Because the *CCC* data is the cumulative count of conforming items between two nonconforming ones, the ARL of the *EWMA*-$X^{1/4}$ for *CCC* data should be calculated as the ARL of the *EWMA* of the transformed data times the mean of the *CCC* data. Some numerical values will be presented in the next section.

For a process shift δ, the ARL can be derived using the mean, variance approximation (Tang, 2000). After using the double square root transformation, the mean of the transformed *CCC* data can be computed.

When the fraction nonconforming of the process is p_0, the mean and standard deviation of the transformed *CCC* data are:

$$E(X_0{}^{0.25}) = \frac{p_0}{1-p_0}[\ln(1-p_0)^{-1}]^{-1.25}\Gamma(1.25) \tag{7.30}$$

and

$$\sigma = [E(X_0{}^{0.5}) - (E(X_0{}^{0.25}))^2]^{0.5} \tag{7.31}$$

where

$$E(X_0{}^{0.5}) \approx \frac{p_0}{1-p_0}[\ln(1-p_1)^{-1}]^{-1.5}\Gamma(1.5) \tag{7.32}$$

When the process fraction nonconforming shifts to p_1, we have:

$$E(X_1{}^{0.25}) = \frac{p_1}{1-p_1}[\ln(1-p_1)^{-1}]^{-1.25}\Gamma(1.25) \tag{7.33}$$

Hence, the shift δ can be obtained as:

$$\delta = \frac{E(X_1{}^{0.25}) - E(X_0{}^{0.25})}{\sigma} \tag{7.34}$$

The values of θ and L should be determined based on the studies of the average run length properties of the *EWMA* control chart. Lucas and Saccucci (1990) provided average run length tables or graphs and the magnitude of the values of θ and L. For a specified $ARL1$ and shift δ in the process, one can easily find the θ and L that produce the minimum $ARL2$.

This *EWMA CCC* chart will be compared with the traditional *CCC* control chart. Table 7.6 represents the sensitivity analysis of two control schemes when p increases. Table 7.7 represents the sensitivity analysis of two control schemes when p decreases.

Table 7.6. The comparison of average run length between the *CCC* chart and *EWMA*- $X^{1/4}$ scheme when the process is deteriorated ($p_0 = 0.0001$)

p	*CCC*	*EWMA*- $X^{1/4}$
0.0001	3855891	3790920
0.00012	4488535	1145467
0.00014	3876941	429329
0.00016	3092498	233913
0.00018	2468115	153517
0.0002	2004505	111690
0.0003	893168	43453
0.0005	322717	17746
0.0008	126754	8833
0.001	81420	6499

Table 7.7. The comparison of average run length between the *CCC* chart and *EWMA*- $X^{1/4}$ scheme when the process is improved ($p_0 = 0.0001$)

p	*CCC*	*EWMA*- $X^{1/4}$
0.0001	3855891	3790920
0.00008	2074256	1232588
0.00006	860242	462633
0.00004	368872	299125
0.00002	213245	284500
0.00001	223008	361300
0.000008	242178	402875
0.000006	278593	470333
0.000005	309500	519000

will be discussed in the next chapter. Run rules could also be considered. On the other hand, there is a need to further investigate the chart properties and the implementation issues.

Another issue that could be of practical interest is the procedures for variable sampling frequency or sampling interval which has attracted some attention for other type of charts recently. Some references are Bai and Lee (1998), Amin and Widmaier (1999), Reynolds and Arnold (2001) and Aparisi and Haro (2001). Similar ideas could be useful for *CCC* chart as well.

Many of the issues raised here are also relevant to the case of Cumulative Quantity Control (*CQC*) chart as described in Section 3.6 (see also Chan *et al.* 2000). It would be of interest to investigate some of the properties of modified *CQC* chart in a similar manner as we have done for *CCC* chart in this chapter. On the other hand, *CQC* chart is based on exponential distribution which is a continuous distribution.

As expected, the *ARL* of the *EWMA*- $X^{1/4}$ scheme decreases more rapidly than the *ARL* of the *EWMA* chart does when the nonconforming fraction p increases. It shows that when the high quality process has deteriorated, the *EWMA*- $X^{1/4}$ control scheme is more sensitive in detecting the shift than the *EWMA* chart. *EWMA*- $X^{1/4}$ control scheme is useful in that it makes it possible for one to detect the assignable cause earlier when the high quality process has deteriorated. Similarly, the *EWMA*- $X^{1/4}$ chart also has good capability in detecting process improvement.

CHAPTER 8

Monitoring of Multiple Process Characteristics

Many industrial processes are multivariate in nature since the quality of a product depends on more than one variable. Multivariate control procedures can be used to capture the relationship between the variables and to provide better process monitoring than that provided by the application of univariate control procedures on each variable. This is an important issue in a high-quality environment as almost all items will be conforming with respect to single attribute. Much work has been done on the multivariate variable processes. However, little attention has been paid to deal with the control of multivariate attribute processes, which is very important in practical production processes.

This chapter deals with multivariate attribute processes with two kinds of procedures: (i) the control chart uses the weighted sum of the counts of nonconforming units with respect to all the quality characteristics as the plotted statistics, and (ii) the control chart uses the multinomial distribution as a probability model for the monitoring of multiple quality characteristics. These charts also enhance the efficiency of identifying the critical assignable cause when an out-of-control signal appears, which is needed when an alarm is raised. Some further research issues are also discussed at the end of this chapter.

8.1. Need and approaches for multiple characteristics

In this section, some general issues of the process control when there are multiple quality characteristics are discussed. We briefly describe the Hotelling's T^2 and then the need for developing methods for multivariate attribute data is discussed.

8.1.1. Need for multiple quality characteristics

With modern data-acquisition equipment and on-line computers it is now a common practice to monitor several quality characteristics simultaneously rather than a single quality characteristic. For example, a chemical process may be a function of temperature, pressure, and flow-rates, all of which need to be monitored carefully. Another example is that an item can be regarded as nonconforming if its width, length or weight does not meet the specification. The former process could be called as a multivariate variable process and the latter one as a multivariate attribute process.

The use of multiple characteristics is important for high-quality products. In a near ZD environment, almost all items are conforming, especially when a single quality characteristic is concerned. When considering more characteristics, the products can be better differentiated, and areas for improvement can also be easily identified. It is in some cases vital to obtain more information about the product and process, and simple methods for monitoring of multiple process characteristics are also very important in automated manufacturing environment.

Processes with multivariate quality characteristics could be monitored by applying a univariate Shewhart chart to each quality characteristics. If the quality characteristics were independent of each other, this would be an adequate procedure. However, multivariate quality characteristics in many production processes tend to be correlated and therefore the results of the univariate chart could be misleading and difficult to interpret.

Montgomery and Mastrangelo (1991) showed that the use of univariate Shewhart charts in a multivariate quality process could distort the simultaneous monitoring of the characteristics, in that the Type I error and the probability of an "in control" point correctly plotted are not equal to their respective levels. A multivariate control procedure would be necessary, as it takes into account the relationship of the correlated quality characteristics. Furthermore, it is more practical and economical to use a single multivariate control scheme rather than several univariate Shewhart control charts. A multivariate control scheme has a better sensitivity than one based on univariate control charts.

Hotelling developed the T^2 procedure for multivariate statistical control, which is by far the most widely discussed. Although we will not focus on this here, a brief description is given.

The Hotelling's T^2 methods is based on a single statistic defined as follows:

$$T^2 = n(\overline{\mathbf{x}} - \overline{\overline{\mathbf{x}}})^t \cdot \mathbf{S}^{-1} \cdot (\overline{\mathbf{x}} - \overline{\overline{\mathbf{x}}}) \tag{8.1}$$

where $\overline{\mathbf{x}}$ is the vector of the mean of individual characteristics and $\mathbf{S}$ is the covariance matrix. This is similar to the traditional χ^2 statistic

$$\chi^2 = n(\overline{\mathbf{x}} - \boldsymbol{\mu})^t \cdot \Sigma^{-1} \cdot (\overline{\mathbf{x}} - \boldsymbol{\mu}) \tag{8.2}$$

Although Hotelling's T^2 does not exactly follow the χ^2 distribution, it can be used to obtain the approximate control limits for a given sample size and an acceptable false alarm probability.

When using simple multivariate control chart, the multivariate data is converted to a single value. The identification of out-of-control characteristic is usually a problem. For more detailed discussion, see standard works such as Montgomery (2001). An extensive review of the multivariate process control is presented in MacGregor and Kourti (1995) in which various multivariate methods are identified along with their applications. A number of basic problems pertaining to multivariate statistical process control are addressed in Mason and Young (1998). Chou *et al.* (1999) considered several alternative covariance estimators and present power determinations for the corresponding T^2 statistic. Some other discussions can be found in Wierda (1994), Liu (1995), Lowery and Montgomery (1995), Spiring and Cheng (1998), Nedumaran and Pignatiello (1999), and Aparisi and Haro (2001).

8.1.2. Monitoring of multiple attribute characteristics

Most of the results on multivariate control charts deal with multivariate variable processes. Although multivariate variable charts are useful in the monitoring of processes, it might be costly to measure all quality characteristics on a continuous scale. In many practical situations, one can easily classify each item inspected as either conforming or nonconforming to the specifications on a given quality characteristic.

Patel (1973) proposed a procedure using the chi-square distribution as an approximation to multivariate binomial and Poisson distributions under the assumptions of time dependence and independence. Marcucci (1985) discussed the use of chi-square control chart when more than one category could be formed within a variable itself. That is, if a nonconforming unit can be classified as major, minor, incidental and none then a chi-square can be employed.

In the following, some multivariate charts for attributes will be discussed in detail. A multivariate chart based on the weighted averages is introduced in Section 8.2. Section 8.3 provides a general multivariate chart based on the multinomial distribution.

8.2. Multivariate *np* (*Mnp*) by weighted average

In the case of multivariate attribute processes, the count of nonconforming units of each quality characteristic follows a Binomial distribution. Considering this, Lu *et al* (1998) proposed a multivariate technique called *Mnp* control chart or multivariate *np*-chart. It uses the weighted sum of the counts of nonconforming units of each quality characteristic in a sample.

8.2.1. The model and control limits

Suppose that there are m quality characteristics. Denote by p_i the probability that an item is nonconforming with respect to quality characteristic i. The quality characteristics may not be independent, and we denote the correlation coefficient between characteristics i and j by δ_{ij}. Note that $\delta_{ij} = \delta_{ji}$ for $i \neq j$ and $\delta_{ij} = 1$, for $i = j$.

Let $\boldsymbol{p}=(p_1,p_2,\ldots,p_m)$ be the fraction nonconforming vector and $\Sigma=[\delta_{ij}]_{mxm}$ be the correlation coefficient matrix. Denote by $\boldsymbol{c}=(c_1,c_2, \ldots,c_m)$ the vector of counts of nonconforming units, where c_i is the count of nonconforming units with respect to quality characteristic i in a sample. A statistic X, which is the weighted sum of the nonconforming units of all the quality characteristics in a sample is defined as

$$X = \sum_{i=1}^{m} c_i / \sqrt{p_i} \tag{8.3}$$

The quality characteristics of a multivariate attribute process affect the process differently even if these characteristics change by the same amount. More specifically, the smaller the fraction nonconforming p_i, the more the count c_i contributes to the statistic X of the multivariate attribute process. In other words, the change of the statistic X is more sensitive to the change of the count c_i with a smaller fraction nonconforming. Therefore, it is obvious that the weight of count c_i must be in inverse proportion to the fraction nonconforming p_i. The reasons for the choice of $1/p_i$ as the weight of quality characteristic i's contribution to test statistic X are as described as follows.

The test statistic X of the *Mnp* chart is a straightforward extension of that of the univariate *np*-chart. Furthermore, the expected value of the test statistic $E(X) = n\Sigma p_i$ is also an increasing function of p_i, which is related to the properties of the process being monitored and agrees with its definition as the weighted sum of nonconforming units with respect to all quality characteristics. Finally, this particular choice makes the derivation of the control limits and sample size requirement easy and presents a clearly defined *Mnp* chart.

It should be noted that the nonconformance of the quality characteristics might carry different physical severity. For example, the nonconformance in one dimension may be physically more serious than that in another dimension. A possible extension is to consider the cost or demerits of nonconformance (Jones *et al.*, 1999). Denoting d_i the number of demerits which indicates the severity of the nonconformance in quality characteristic i, the above test statistic X can be extended to be more realistic in the following manner:

$$X_D = \sum_{i=1}^{m} d_i c_i / \sqrt{p_i} \tag{8.4}$$

where $D = (d_1, d_2, ..., d_m)$ is the nonconformance vector. The mean and variance of the test statistic X can be calculated as follows.

$$E(X_D) = \sum_{j=1}^{m} d_j n p_j / \sqrt{p_j} = n \sum_{j=1}^{m} d_j \sqrt{p_j} \tag{8.5}$$

and

$$Var(X_D) = n\left(\sum d_j^2 (1 - p_j) + 2 \sum_{i<j} d_i d_j \delta_{ij} \sqrt{(1 - p_i)(1 - p_j)} \right) \tag{8.6}$$

The correlation coefficient δ_{ij} can be either known or estimated with the following equation from the preliminary samples:

$$\delta_{ij} = Cov(c_i, c_j) / \sqrt{Var(c_i)Var(c_j)}$$
$$= E\left[(c_i - E(c_i))(c_j - E(c_j))\right] / \sqrt{Var(c_i)Var(c_j)}$$

Using the general Shewhart control chart principles, we can obtain the control limits for the Shewhart-type *Mnp* chart. If the mean and variance of the statistic X are μ_X and σ_X respectively, then the general model for the Shewhart control chart is:

$$\begin{aligned} UCL &= \mu_X + k\sigma_X \\ CL &= \mu_X \\ LCL &= \mu_X - k\sigma_X \end{aligned} \tag{8.7}$$

where k is the distance of the control limits from the central line, in multiples of the standard deviation of X. It is customary to use k=3; and the corresponding false alarm probability 0.0027 is considered acceptable.

Assume that the fraction nonconforming vector $\boldsymbol{p}=(p_1,p_2,\ldots,p_m)$ and the correlation matrix Σ of a multivariate attribute process are known or have been estimated, the central line and control limits of the Shewhart-type *Mnp* control chart are

$$UCL = n\sum_{j=1}^{m} d_j\sqrt{p_j} + 3\sigma_m \tag{8.8}$$

and

$$LCL = n\sum_{j=1}^{m} d_j\sqrt{p_j} - 3\sigma_m \tag{8.9}$$

where

$$\sigma_m = \sqrt{n\left\{\sum_{j=1}^{m} d_j(1-p_j) + 2\sum_{i<j}(d_i d_j \delta_{ij}\sqrt{(1-p_i)(1-p_j)})\right\}}$$

Given the values of the parameters, the control limits can be computed and the *Mnp* chart can then be established. It is noted that with equal nonconformance severity in m quality characteristics, the limits could be computed with d_i=1 for all i.

8.2.2. The implementation

When implementing the procedure, several issues need to be resolved. Some are discussed in the following.

Estimation of model parameters

When the process fraction nonconforming vector $\boldsymbol{p}$ and the correlation matrix Σ are not known, they must be estimated from observed data. We can first take k preliminary samples, each with size n. As a general rule of statistical process control, k should be more than 25. Let c_{ij}, i=1,...,m; j=1,...,k denote the nonconforming units of quality characteristic i in sample j. The fraction nonconforming vector of sample j, $\boldsymbol{p}_j$, is estimated as:

$$\begin{aligned}\overline{\boldsymbol{p}}_j &= \left(\frac{c_{1j}}{n}, \frac{c_{2j}}{n}, \ldots, \frac{c_{mj}}{n}\right) \\ &= (\overline{p}_{1j}, \overline{p}_{2j}, \ldots, \overline{p}_{mj}), \quad j = 1,2,\ldots,k\end{aligned} \tag{8.10}$$

and the overall fraction nonconforming vector $\boldsymbol{p}$ is estimated as

$$\begin{aligned}\overline{\boldsymbol{p}} &= (\overline{p}_1, \overline{p}_2, \ldots, \overline{p}_m) \\ &= \frac{\sum_{j=1}^{k} \overline{\boldsymbol{p}}_j}{k} = \left(\frac{\sum_{j=1}^{k} c_{1j}}{nk}, \frac{\sum_{j=1}^{k} c_{2j}}{nk}, \ldots, \frac{\sum_{j=1}^{k} c_{mj}}{nk}\right)\end{aligned} \tag{8.11}$$

The correlation matrix Σ is also estimated from the preliminary samples:

$$\delta_{ij} = \frac{Cov(c_i, c_j)}{\sqrt{Var(c_i)Var(c_j)}} \tag{8.12}$$

Using the above, the limits are determined as

$$UCL = n\sum_{j=1}^{m} \sqrt{\overline{p}_j} + 3\hat{\sigma}_m \tag{8.13}$$

and

$$UCL = n\sum_{j=1}^{m}\sqrt{\bar{p}_j} - 3\hat{\sigma}_m \tag{8.14}$$

where

$$\hat{\sigma}_m = \sqrt{n\left\{\sum_{j=1}^{m}(1-\bar{p}_j) + 2\sum_{i<j}(\bar{\delta}_{ij}\sqrt{(1-\bar{p}_i)(1-\bar{p}_j)})\right\}}$$

We regard the control limits obtained above as trial control limits. They allow us to determine whether the process is in statistical control when k initial samples are selected. To test the hypothesis, the sample statistic X can be plotted from each sample on the *Mnp* chart. If all points are within the control limits, then we can conclude that the multivariate attribute process has been in control, and the trial control limits are suitable for controlling current or future production. Otherwise assignable causes should be found and control limits can be revised.

Interpreting out-of-control signals

The most difficult part of applying a multivariate control chart is the interpretation of out-of-control signals (Doganaksoy *et al.*, 1991; Hayter and Tsui, 1994 and Runger *et al.*, 1996). For example, if a multivariate chart shows an out-of-control signal due to the joint effect of two or more correlated continuous characteristics, the univariate charts on the separate continuous characteristics may not even show as signal and it is difficult to tell the cause of the alarm. This topic has been discussed by a number of authors. Methods such as monitoring the individual continuous characteristics or the principal components with univariate charts, examining the individual continuous characteristics using Bonferroni limits for the univariate charts have been proposed.

Wade and Woodall (1993) used regression adjustments for individual continuous characteristics to improve the diagnostic power of multivariate Hotelling's T^2 charts following a chart signal. Mason *et al.*

(1995) decomposed the T^2-statistic into individual components each of which provides information on the continuous characteristics that significantly contribute to an out-of-control signal.

There is a simple approach to interpreting an out-of-control signal for an *Mnp* chart using a score statistic. A statistic X based on the weighted sum of the counts of nonconforming units of all the quality characteristics can be used. The weight of quality characteristic i is a decreasing function of the fraction nonconforming of quality characteristic i. This procedure indicates that the smaller the fraction nonconforming p_i, the more the change of quality characteristic i contributes to the shift of statistic X. Therefore the score statistic which is defined as

$$Z_i = [c_i - np_i] / \sqrt{\overline{p}_i} \tag{8.15}$$

and

$$Z_{Di} = d_i [c_i - np_i] / \sqrt{\overline{p}_i} \tag{8.16}$$

when nonconformance severity is not equal for different quality characteristics can be used.

At an out-of-control signal, we can first compute the score of every individual quality characteristic. Without loss of generality, we assume that Z_j and Z_j give the smallest score and the largest score respectively. If the statistic X falls above the upper control limit, Z_j gives the largest positive score and quality characteristic j is considered the critical contributor to the upward shift of the process. On the other hand, if the statistic X falls below the lower control limit, Z_i gives the smallest negative score and quality characteristic i is considered as the critical contributor to the downward shift of the process.

Choice of sample size

As for any other control chart an appropriate sample size n should be chosen for effective application. If the fraction nonconforming vector p is very small, the sample size n should be sufficiently large so that we have a high probability of finding at least one nonconforming unit of a quality characteristic in the sample. Otherwise we might find that the control limits are such that the presence of only one non-conforming unit in the sample would indicate an out-of-control condition.

To avoid this undesirable situation, we can choose the sample size n so that the probability of finding at least one nonconforming unit per sample is at least equal to some specified value (such as 0.95). Using the cumulative Poisson table in a way similar to univariate *np*-chart procedures, the condition for the sample size of the *Mnp* chart is

$$n\sum_{i} p_i \geq 3m \tag{8.17}$$

If the in-control value of the fraction nonconforming vector p is small, we can also choose n large enough so that the *Mnp* control chart will have a positive lower control limit. This ensures that we will have a mechanism to trigger an investigation of one or more samples that contain an unusually small number of nonconforming units. Under the matrix notation, since we wish to have:

$$LCL = n\sqrt{\bar{\boldsymbol{p}}\mathbf{1}'} - 3\sqrt{nTrace(\sqrt{(\mathbf{1}-\bar{\boldsymbol{p}})(\mathbf{1}-\bar{\boldsymbol{p}})'\Sigma'}} > 0 \tag{8.18}$$

this implies that

$$n > 9\frac{Trace(\sqrt{(\mathbf{1}-\bar{\boldsymbol{p}})(\mathbf{1}-\bar{\boldsymbol{p}})'\Sigma'})}{(\sqrt{\bar{\boldsymbol{p}}\mathbf{1}'})^2} \tag{8.19}$$

where $\mathbf{1}= [1,1,\dots,1]$ is the unit vector.

With the trial control limits, the *Mnp* chart could be easily plotted. Assume that ten additional samples are taken and the data are shown in Table 8.1 itself from the sample number twenty-one onwards. The *Mnp* chart is plotted in Figure 8.1, in which the statistic *X* is plotted against the sample sequence. From the chart, it is obvious that all the first 20 points are within the control limits and also no systematic patterns are found and so the process is considered in statistical control. Therefore, it is reasonable to use the trial limits as control limits for monitoring this multivariate attribute process.

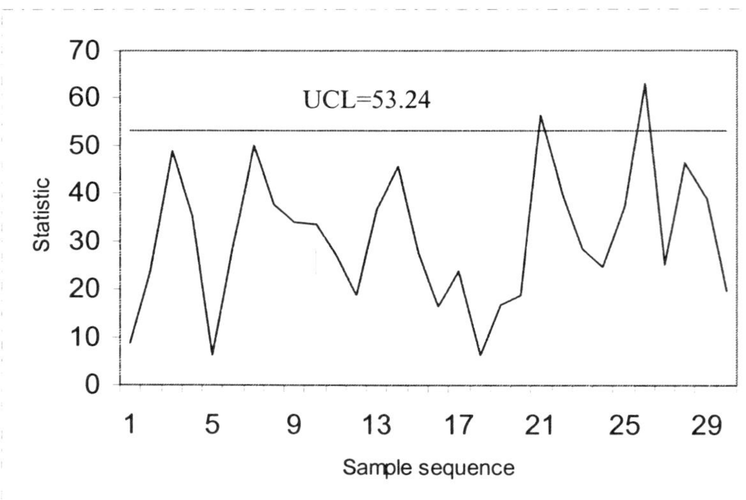

Figure 8.1. The *Mnp* chart for the simulated data in Table 8.1

From the Figure 8.1, we can see from sample 21, the process seems to have shifted to a high level. The points 21 and 26 are out of the upper control limit, so it is clear that the nonconforming fraction vector of the process has increased. The process should be stopped for an investigation and possible assignable causes for the process deterioration should be found.

A follow-up question is to identify the critical one that has contributed most to the out-of-control signal in the multivariate chart. The three scores for sample 21 and 26 are $z_1 = 14.73$, $z_2 = 3.93$, $z_3 = 10.55$ and $z_1 = 14.73$, $z_2 = 10.48$, $z_3 = 10.55$, respectively. Note that the score z_1 is the largest for both sample 21 and sample 26. Hence, the first quality characteristic is probably the critical contributor to process deterioration. Note that the individual *np*-chart for characteristic 1 in this case shows that the process is in control (Lu *et al.* 2000), so the process shift could be due to some joint causes.

□

8.3. A chart with multinomial distribution

A straightforward method of generalization of binomial distribution is the multinomial distribution, which can be used in constructing multivariate attribute chart. Such a chart called multinomial attribute chart (*MAC*) can take care of correlation by itself by properly defining the quality class of a product. We briefly describe the procedure and discuss the implementation. The proposed chart is easy to apply in modern manufacturing environment whether the variables are attributes or continuous characteristics.

8.3.1. Use of multinomial distribution

Consider a multivariate process with m mutually exclusive quality classes of interest. Each quality class is a combination of quality characteristics being measured on a go and no-go basis in a sample of size n. Let $c_1, c_2, \ldots, c_m$ be the number of units in each quality class. Then the vector $\boldsymbol{c}$ can be modeled with a multinomial distribution, $M(n,p)$ with vector of cell probabilities $p_1, p_2, \ldots, p_m$ of m classes.

Note that this formulation is general. Suppose that there are k quality characteristics, a number of quality classes can be formed depending on the information available. Suppose two characteristics A, B are available. One class could be items with Type A nonconformity, and another class could be items with only Type B nonconformity. A third could be those with both, and finally, there is one with no nonconformity. Another class could also be a combination of the classes or characteristics depending on the need. In fact, when nonconforming items could be measured for their severity, such as critical defect, major defect, minor defect and no defect, this would be a special case of our model description.

Assume that the observations may have m different results, U_1, U_2, …,U_m, which are mutually exclusive and the probability of the result U_i is p_i such that $p_1 + p_2 + ... + p_m = 1$. Then the probability mass function is given by

$$M(n, p) = n! \prod_{i=1}^{m} \frac{p_i^{x_i}}{c_i!} \quad (8.20)$$

which denotes the probability that the event U_1 will occur c_1 times, the event U_2 will occur c_2 times, …, and the event U_m will occur c_m times in n observations such that $c_1 + c_2 + ... + c_m = n$.

Consider now n multinomial trials with m possible mutually exclusive outcomes namely c_i (i = 1,2,…,m). Suppose it is desired to test a simple hypothesis concerning the goodness of fit between the observed and expected frequencies. The well-known chi-square statistic can be used, with

$$X^2 = \sum_{1}^{m} (x_i - np_i)^2 / np_i \quad (8.21)$$

The statistic would be approximately distributed as chi-square with m-1 degrees of freedom subject to correlation between the c's and the relationship $\sum c_i = n$. With the c's mutually independent, Eq. (8.21) can be replaced by

$$X^2 = \sum_{1}^{m} (x_i - np_i)^2 / np_i(1 - p_i) \tag{8.22}$$

which is chi-square distributed with m degrees of freedom.

From the above equations, the null hypothesis is whether the intended process meets the requirements of desired process averages. This is equivalent to test $H_0 : \overline{p} = \overline{p}_0$ versus $H_1 : \overline{p} \neq \overline{p}_0$, where $\overline{p}$ is the vector of m variables and $\overline{p}_0$ is the desired process averages. Under H_0, the process is considered as in-control almost (1-α) percent of the time. That is, using chi-square distribution, the decision would be taken that the process is in control if the test statistic $X^2 \leq \chi^2(m\text{-}1,\alpha)$ and the process is out of control if $X^2 > \chi^2(m\text{-}1,\alpha)$. The test statistic under the alternative hypothesis does not follow the chi-square distribution. Instead, it follows a noncentral chi-square distribution (Johnson and Kotz, 1970) with m-1 degrees of freedom with noncentrality parameter λ.

Example 8.2:

Consider an industrial process with k quality characteristics. For example, a process goes through k operational stages consecutively. At the end of the k:th operation, a certain measurement is taken. Now, let p_i be the probability of being nonconforming of a characteristic i. Let $p_{i,j...l}$ be the probability of being nonconforming unit with $i,j,...,l$ characteristics at the same time. That is

$$p_{i,j,...,l} = p_{i<j<...<l}(U_i, U_{j,...,} U_l),\ \ i,j,l \in \{1,2,\ldots,m\}$$

If we assume p_c be the probability of being a conforming unit then we write

$$p_c = 1 - \sum_{i=1}^{m} p_i - \sum_{i<j<...<l}^{m} p_{i,j,...,l}$$

Now the multinomial distribution would be $M(n;p_1,p_2,..,p_m,p_g,p_c)$ where

$$p_g = \sum_{i<j<\ldots<l}^{m} p_{i,j,\ldots,l}$$

and hereafter we call it a general characteristic.

For instance, assume the process average as 0.002 (p_1), 0.0025 (p_2) and 0.001 (p_3), respectively, for three characteristics. If we take p_g=0.004 as a general characteristic, then p_c= 0.9905. This is an example of five classes.

However, extending this example for the case of more categories within a single attribute such as major, moderate, minor and none, a multinomial process can still be assumed as suggested by Marcucci (1985). For example, as suggested by Nelson (1987), carbon resistors classed as 1%, 5%, 10% and >10%; or a product graded into A, B, C and D classes. That is, assuming the probability of being a nonconforming unit is 0.01 then this can be further classified as 0.001, 0.005 and 0.004 for major, minor and none respectively. Then the multinomial distribution would be $M(n;0.001,0.005,0.004,0.99)$. It is noted that the interactions between these classes have been ignored. Thus, it will lead a simple chi-square chart or a summary chart only, not *MAC*. However, it can be treated as a special case of *MAC* assuming there is no interaction between the classes.

□

Estimation of parameters

Under the proposed procedures, the number of parameters for the k characteristics in this *MAC* chart is greater than m since the interactions between the characteristics are considered. It is noted that m is not necessarily be greater than k as still some of the characteristics can be ignored while grouping. For instance, two chracteristics, say A and B, can be grouped as A and (B+AB). Then the parameters could be estimated as follows. Let x_i be the number of nonconforming items of characteristic i in a sample of size n, then an estimate of p_i is

$$\hat{p}_i = \frac{x_i}{n}$$

Similarly, let $x_{i,j,\ldots,l}$ be the number of nonconforming items of characteristic $i, j, \ldots, l$, in a sample of size n, then an estimate of $p_{i,j,\ldots,l}$ is

$$\hat{p}_{i,j,\ldots,l} = \frac{x_{i,j,\ldots,l}}{n}$$

and

$$\hat{p}_c = 1 - \sum \hat{p}_i - \sum \hat{p}_{i,j,\ldots,l}$$

It is noted that assuming either $p_{i,j,\ldots,l}$=0 or merger with p_c will cause serious problems when the attributes are really correlated. However, for prefixed (known) p_i's, the problem of estimation is not easy as discussed above and will not be discussed here.

Example 8.3:
Suppose a process of two types of quality characteristics is considered with the conditional process averages of p_1=0.001 and p_2=0.003. Now assume that the probability of occurrence of the general characteristic, that is two characteristics together, in an item is p_g=0.002. Taking a sample size of 5000, the multinomial parameters are M(5000;0.001,0.003,0.002,0.994), where 0.994 is p_c, the probability of obtaining a defect free item. Under this multinomial assumption, Table 8.2 has been generated using Kemp and Kemp's method, see Johnson *et al.* (1996), in which the last twenty-five samples are with shifted process average. This process shift was introduced after 100th sample, from 0.001 to 0.0015 in characteristic 1 alone.

Assuming a false alarm probability α=0.005 with the chi-square value of 12.8381 (for 3 degrees of freedom), a control chart has been drawn using 8.21 in Fig.8.2.

Table 8.2. A simulated data of multivariate processes for a sample of size 5000 with three process averages

p	0.001	0.003	0.002	0.9940	χ^2
Sample	Count	Count	Count	Count	
1	5	25	16	4954	10.38
2	7	18	17	4958	5.82
3	7	8	7	4978	4.98
4	1	13	9	4977	3.52
5	1	20	5	4974	7.94
6	5	18	14	4963	2.02
7	8	14	11	4967	1.96
8	5	12	9	4974	0.66
9	8	14	12	4966	2.17
10	6	8	10	4976	3.24
...	...	...	...	...	...
96	7	10	9	4974	2.51
97	7	8	11	4974	3.89
98	2	17	11	4970	2.17
99	4	20	9	4967	2.35
100	0	16	9	4975	5.27
101	12	20	14	4954	13.12
102	12	18	11	4959	10.52
103	8	15	9	4968	1.90
104	11	15	6	4968	8.80
105	1	13	11	4975	3.57
...	...	...	...	...	...
121	5	11	7	4977	1.98
122	12	9	13	4966	13.10
123	9	10	12	4969	5.27
124	9	17	9	4965	3.57
125	6	16	10	4968	0.27

The chart shows an out of control signal at the 101st sample and keeps on showing the signals at different intervals about 4 times within 25 shifted samples. This is sufficient to show the process shifts clearly.

□

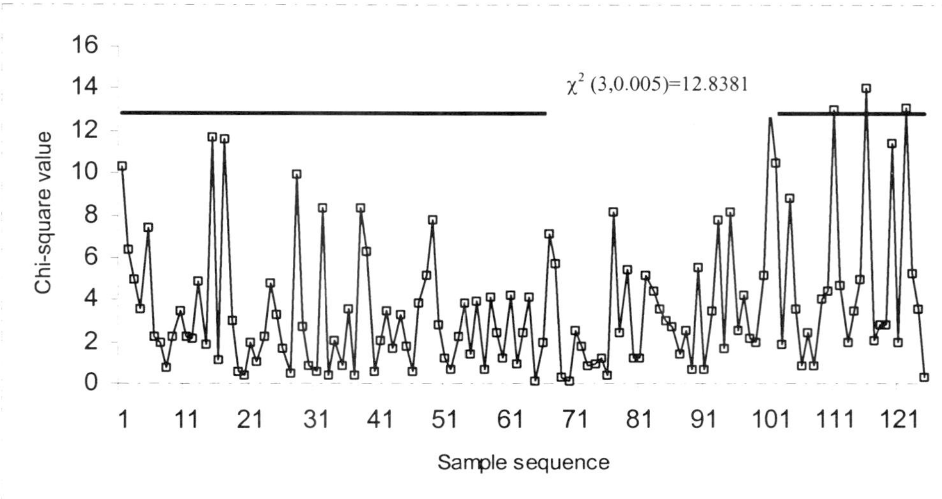

Figure 8.2. A multivariate control chart for the count data with p_1=0.001, p_2=0.003, p_g=0.002 and p_0=0.994

8.3.2. A study of Average Run Length (*ARL*)

The efficiency of any Shewhart control chart can be measured using the average run length distribution, which means the average length of conforming units between two out of control signals. Without loss of generality, let β be the probability of in-control process when the process average is p. We have that under H_0,

$$\beta = P[X^2 \le \chi^2(m-1,\alpha)] \tag{8.23}$$

As in the univariate chart, the process shift can be detected for any specific alternative hypothesis or in particular some other desired p values, though normally this may not be the case. Suppose that under the alternative hypothesis, the outcome i has the probability $\overline{p}^*$, that is $H_1 : \overline{p}^* = \overline{p}$, then the power of the test is given by

$$1-\beta = P[\lambda \geq \chi^2(m-1)] \tag{8.24}$$

where

$$\lambda = n\sum_{i=1}^{m}(p_i^* - p_i)^2 / p_i \tag{8.25}$$

and $\chi^2(m)$ is the noncentral chi-square distribution and λ is the noncentrality parameter, which is simply the distance between $\overline{p}$ and $\overline{p}^*$. Then the average run length is calculated by

$$ARL = 1/(1-\beta) \tag{8.26}$$

Note that the *ARL* value for in-control process average is always $1/\alpha$ since λ becomes zero. This implies that on the average one would expect a false alarm rate at every $(1/\alpha)^{\text{th}}$ observation. On the other hand, when the process has shifted to some other value say, p^*, then the noncentral chi-square should be used to obtain the *ARL* with noncentrality parameter λ. The *ARL* can also be used to fix the sample size and sampling frequency. This means for any desired *ARL*, sample size can be fixed.

Example 8.4:

In order to demonstrate the *ARL* properties, assume that the process is shifted in variable 1 from 0.001 to some other value. The power of tests and *ARL* values are calculated from noncentral chi-square tables using non-centrality parameter λ with α= 0.005 as shown in Table 8.3.

Table 8.3 indicates that the detection of process changes seems to be good. For example, assume the process mean average is shifted from 0.001 to 0.002 in variable 1. From Table 8.3, it requires 6 samples to detect the shift while the maximum true *ARL* is maintained at 200. Similarly for process improvement, suppose the mean is shifted to 0.0001, then it takes 9 samples.

□

Table 8.3. *ARL* values for shifted process averages in p_1 when for α= 0.005

p_1	p_2	p_g	p_c	λ	$1-\beta$	*ARL*
0.0001	0.003	0.002	0.994	4.0500	0.11589	8.63
0.0002	0.003	0.002	0.994	3.2000	0.08099	12.35
0.0004	0.003	0.002	0.994	1.8000	0.03662	27.31
0.0006	0.003	0.002	0.994	0.8000	0.01559	64.14
0.0008	0.003	0.002	0.994	0.2000	0.00715	139.80
0.0010	0.003	0.002	0.994	0.0000	0.0050	200.00
0.0012	0.003	0.002	0.994	0.2000	0.00715	139.80
0.0014	0.003	0.002	0.994	0.8000	0.01559	64.14
0.0016	0.003	0.002	0.994	1.8000	0.03662	27.31
0.0018	0.003	0.002	0.994	3.2000	0.08099	12.35
0.0020	0.003	0.002	0.994	5.0000	0.16102	6.21

A good control scheme must provide good sensitivity in terms of low *ARL* when an undesirable factor is affecting the process and high *ARL* when the process is stable. Some *ARL* curves for different number of quality characteristics are shown in Figure 8.3.

8.3.3. Some practical issues

There are several practical issues when implementing multivariate chart in practice. Some of them are briefly discussed here. Although this part is related to *MAC* chart, many of the issues are common for all multivariate charts.

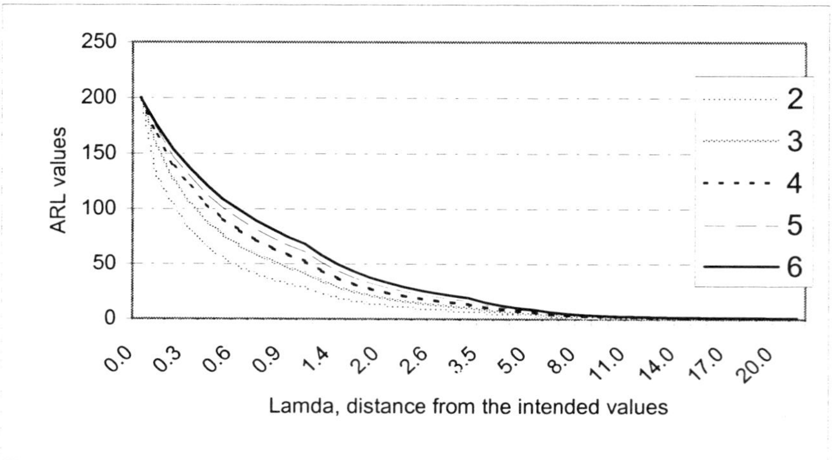

Figure 8.3. Curves of *ARL* when attributes are increased in a multivariate procedure.

Sample size determination

Determination of the optimum sample size is very important. It involves a number of factors such as time, inspection cost, storage cost, etc. For a given *ARL*, the sample size can be found using the following steps:

(1) Calculate the power of the test as $1-\beta=1/ARL$;

(2) Obtain the value of noncentral parameter (λ) using the power of the test with Eq. (8.24) for a given false alarm probability (α);

(3) Obtain the sample size from Eq. (8.26) for a given process average:

$$n=[\frac{\lambda}{\sum(p_i^*-p_i)^2/p_i}] \qquad (8.27)$$

Example 8.5:
To examine the issue of sample size, we first obtain the *ARL* value for the worst process shift in order to safeguard against extreme quality deterioration. For instance, let the shift takes place from 0.001 to 0.0014 for the first characteristic and the others remain the same at 0.003, 0.003, 0.002 and 0.994. Assume that the maximum *ARL* is 25. The sample size is then found using Eq.(8.27) as 10000 for the false alarm probability of 0.005. Suppose that the *ARL*s 16 and 52 are desired, the corresponding sample sizes are 13750 and 5000, respectively.

For any given *ARL*, the sample size can be found and vice-versa. A set of values of λ and *ARL* are presented in Table 8.4 for 2-4 classes.

□

Interpretation of out of control signal
The main problem in using χ^2 chart is the interpretation of out of control signal since there are m quality classes. Among the m quality classes, it is of our interest to find out the exact quality classes that causes the process to be out of control. Unless this is identified, the signal could not be diagnosed easily. Unlike the multivariate normal distribution, chi-square control chart is easy to use when there is a shift in the variable. The reason is that it could be easily decomposed into m χ^2 statistics such as

$$\chi^2 = z_1 + z_2 + \ldots + z_m \tag{8.28}$$

where

$$z_i = (c_i - np_i)^2 / np_i$$

Table 8.4. *ARL* values corresponding to non-centrality parameter (λ)

λ	# classes			# classes		
	2	3	4	2	3	4
	ARL values for α=0.5%			*ARL* values for α=0.1%		
0	200	200	200	1000	1000	1000
0.1	128	156	167	625	714	769
0.2	103	127	139	435	556	625
0.4	66	89	104	250	370	435
0.6	47	67	81	167	256	323
0.8	36	52	64	120	189	244
1	28	42	52	91	145	192
1.2	23	34	44	71	115	154
1.4	19	29	37	57	93	125
1.6	16	25	32	47	77	103
1.8	14	21	27	39	64	87
2	12	18	24	33	54	74
2.2	11	16	21	28	46	63
2.4	10	14	19	25	40	55
2.6	9	13	17	21	35	48
2.8	8	12	15	19	31	42
3	7	11	14	17	27	37
3.5	6	8	11	13	20	28
4	5	7	9	10	16	22
4.5	4	6	7	8	13	17
5	4	5	6	7	10	14
6	3	4	5	5	7	10
7	2	3	4	4	6	7
8	2	3	3	3	4	5
9	2	2	3	3	3	4
10	2	2	2	2	3	4

Furthermore, the chi-square statistic is a positive real number and hence, each z_i is a positive real number. Thus each z_i takes a value from zero onwards. If the value of z_i is zero, then the quality classes i is 100 percent in-control. Otherwise, it will have some deviation from the intended value. Among the m statistics, the largest value of z_i would contribute most to the deviation. When the deviation exceeds certain amount, it will show an out of control signal. Suppose we arrange them according to their magnitudes, we could easily identify the corresponding quality classes that caused the process to be out of control. This means that the highest value of z_i may be treated as the main cause for the signal when it occurs.

For example, take the case discussed previously. From the chart (Fig.8.2), the signal is shown at the sample point at 101. The corresponding z_i's are 9.80, 1.60, 1.67 and 0.052 for classes 1, 2, 3 and 4 respectively. The highest value 9.8 falls on class 1. Therefore it can be concluded that the quality class 1 is the main source that lead the process out of control.

8.4. Some issues for further research

Although multivariate process control has attracted a lot of attention recently, most of the results deal with continuous measurement characteristics. Multivariate attribute data has a great potential for research. In fact, when continuous process characteristics are available, they can be easily grouped and some attribute control charts can be used. In Section 8.3, the *MAC* is presented in such a way that it allows general classification of quality characteristics.

An extended issue is the combination of continuous and discrete data. This is a challenging issue and further investigation is needed. Usually variable charts and attribute charts are discussed separately, but it is common for products to have both types of quality characteristics at the same time. The characteristics can very well be correlated and models for this type of processes need to be developed. Of course, it could be possible to classify variable quality characteristics and use multivariate attribute chart, and we may also consider attribute characteristics as variable characteristic and use traditional multivariate chart for variables. However, these are first-order approximations and their accuracy has to be assessed.

For multivariate charts, an important issue is to estimate the process parameters. These parameters include the ones related to the correlation of the characteristics which could be difficult to measure. When the parameters are estimated, their accuracy also has to be assessed.

The investigation of the cause, or the specific quality characteristic that has been the cause, when a single chart shows an out-of-control signal is still an important issue. Although various approaches are available, the cost of wrong decision could be different. Hence, some cost models could be developed for this purpose.

It should be pointed out that the proposed methods for monitoring multivariate attribute data might not be suitable for high-quality manufacturing processes. The problems could be the same for univariate attribute charts. Similar extension might be possible. On the other hand, with stringent requirements, but considering more quality attributes, analysis of high-quality production processes is an alternative and should be considered in practice.

Little has been reported on the economic design of multivariate control charts. This is a related but important issue. In Chapter 9, economic design of control chart is discussed, but again, they are restricted to single quality characteristic. Further discussions on this subject will be given at the end of that chapter.

CHAPTER 9

Economic Design of Geometric Chart

Cost consideration is always important in industry and any quality related activities should be put into the context of cost saving to improve profitability. In using control chart in practice, economic factors can be taken into account for designing control chart. In this chapter, the selection of design parameters is investigated from an economic point of view. Specifically, the sampling interval and the control limit for the *CCC* chart will be studied. An economic model for *CCC* chart design, which is based on the flexible cost model by Lorenzen-Vance, is studied. A simplified procedure is used to derive the optimum solution. Some sensitivity analysis is also presented to investigate the effect of input parameters on the optimum economic design of *CCC* chart.

The economic design of control chart provides a method to optimize the risk of false alarms and that of failing to detect the process shift from an economic point of view. However, in doing so, the statistical properties might be sacrificed. An approach to economic charting that meets statistical properties, i.e. economic-statistical design, is also discussed in this chapter. We will focus on the use of *CCC* chart although the study can be extended to other related control charts. Some other related extensions and topics for further research are mentioned at the end of this chapter.

9.1. The Lorenzen and Vance model for economic design

The design of control charts based on cost consideration involves the development of cost models. A general economic model was presented in Lorenzen and Vance (1986) which is an extension of the earlier model by Duncan (1956). Since the Lorenzen and Vance (LV) model is extensively used in the rest of this chapter, we briefly describe this model here.

Lorenzen and Vance (1986) developed a generalized process model that can be applied to all control charts regardless of the monitoring statistic used. It is only necessary to compute $ARL1$ and $ARL2$ of the control chart in some specified fashion. Any distribution can be used to model the time to the occurrence of an assignable cause. Furthermore, the model incorporates two indicator variables, δ_1 and δ_2. They are used to denote whether the process continues or discontinues during the search for the assignable causes and during process repair stage. That is, $\delta_1 = 1$ if production continues during the search, 0 otherwise; and $\delta_2 = 1$ if production continues during the repair, 0 otherwise.

A production cycle is defined as the time period from the beginning of production or after an adjustment to the detection and elimination of an assignable cause. The cycle time consists of the five parts, $A_1, A_2, \cdots, A_5$, denoting the five segments of the production cycle:

A_1 - the time until the assignable cause occurs;

A_2 - the time until the next sample is taken;

A_3 - the time to analyze and chart the sample;

A_4 - the time until the chart gives an out-of-control signal; and

A_5 - the time to discover the assignable cause and to repair the process.

Figure 9.1 in which the stage A_3 is omitted, illustrates the process.

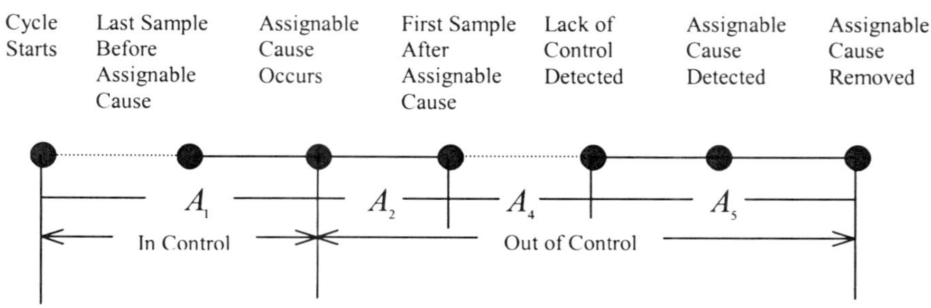

Figure 9.1. Illustration of in-control and out-of-control states of a process.

In his pioneering work in the economic design, Duncan (1956) proposed a model for the design of $\overline{X}$ chart. Many subsequent economic designs of control charts adopted the basic concepts from this initial work, e.g., McWilliams (1994), Surtihadi and Raghavachari (1994) and Chou (2001). The development of the LV cost model closely follows Duncan's single cause, exponential failure function assumption. Instead of using the expected hourly net income, the expected hourly cost function is the optimality criterion. General discussions on economic design can be found in Montgomery (1980) and Ho and Case (1994). It should be pointed out that most of the research on economic design focus on the design of variable charts. Some papers dealing with attribute chart design are Chiu (1975), von Collani (1988), Chung

(1995), Tang *et al.* (2000), Wu *et al.* (2000) and Xie *et al.* (1997, 2001b).

The LV model attempts to minimize the expected hourly production cost. In contrast, Duncan's model tries to maximize expected net income. However, when Duncan searched for an optimal solution, he translated the objective of maximizing the expected net income into minimizing the expected loss. Thus, both models essentially end up with the same objective, that is, to minimize total cost. McWilliams (1996) presented some studies that relate earlier economic models to the LV unified model.

9.2. Economic design of *CCC* chart

All economic design is based on minimizing a kind of expected unit or cycle cost with respect to some design parameters. As discussed before, we will mainly adopt the Lorenzen-Vance approach. The cost function is described. Numerical procedure and examples are presented as well.

9.2.1. The expected cost function

Notations and Assumptions

The approach by Lorenzen and Vance (1986) not only provides the unified approach to economic design of control charts, it also has a unified notation. With some modifications, the following notations can be used in the formulation of the cost function for the *CCC* chart in this chapter.

The design parameters are as follows:

l: (one-sided) control limit for the *CCC* control chart;

h: sampling interval.

The cost and process parameters are as follows:

λ: 1/mean time the process is in control;

E: time to sample and chart one item;

T_0: expected time spent searching for a false alarm;

T_1: expected time spent finding the assignable cause;

T_2: expected time spent rectifying the process;

$\delta_1 = 1$ if the production continues during a search and 0 otherwise;

$\delta_2 = 1$ if the production continues during a repair and 0 otherwise;

C_0: quality cost/hour while the process is in control;

C_1: quality cost/hour while the process is out of control (> C_0);

Y: cost per false alarm;

W: cost to locate and repair the assignable cause;

a: fixed cost per sample;

b: cost per unit sampled;

p_0: expected fraction defective produced when the process is in control;

p_1: expected fraction defective produced when the process is out of control;

s: expected number of samples while in control;

$h-\tau$: expected time between the shift in the process and the next sample;

$ARL1$: average run length when the process is in control;

$ARL2$: average run length when the process is out of control;

$E(C)$: expected cost per hour.

The process is assumed to start in a state of statistical control with a known fraction nonconforming of p_0. There is a single assignable cause and its occurrence results in a change of the fraction nonconforming of the process from p_0 to p_1. The time between occurrences of the assignable causes is exponentially distributed with a mean of λ occurrences per hour. This assumption is commonly used in the economic design, see e.g., Chiu (1975), Montgomery (1982) and Lorenzen and Vance (1986).

Assume that sampling is based on one-at-a-time measurement. Thus the sample size is equal to one. This assumption is not limited in many practical situations and it is usually the case in automated manufacturing environment. Furthermore, measurements made one by one can give a more accurate representation of the inspection process in the usual cases, for instance, in continuous flow processes, see Tagaras (1994). This assumption greatly simplifies the economic design of the *CCC* chart in two aspects. First, since the sample size is always "one", the cost of taking a sample and inspecting it is a known constant, $a+b$. Secondly, the optimal parameters are reduced to two, namely, control limit for the *CCC* control chart l and the sampling interval h.

Only the low-sided *CCC* chart is studied in this chapter because of its simplicity and practicality. When the *CCC* data is beyond the upper control limit, the process is better and the fraction of nonconforming is smaller, which causes no actual problems. Therefore, we are more concerned with the case of getting out of the lower control limit (*LCL*). Actions should be taken in this situation. In the course of inspection, one item is checked every *h* hours of production. The number of consecutive conforming items is counted until a nonconforming one is detected. If the count is larger than or equal to *LCL*, no action is taken. Otherwise, assignable causes should be searched and then eliminated.

The overall cost model

Based on the above notations and assumptions, the loss function for an one-sided *CCC* chart can be formulated as in Section 9.1. Combine the value of A_1 - A_5; we can obtain the expected cycle time for *CCC* chart as follows (Xie *et al.*, 1995):

$$\begin{aligned} E(T) &= \sum_{i=1}^{5} A_i \\ &= \frac{1}{\lambda} + \frac{(1-\delta_1)sT_0}{ARL1} - \tau + E + hARL2 + T_1 + T_2 \end{aligned} \tag{9.1}$$

Note that E is the expected time to sample and chart one item and it is often taken as zero. For *CCC* chart, $A_3 = E$ because the sampling is based on one-item-a-time measurement In fact, this type of measurement is more and more widely used as automation spreads over the manufacturing industry.

Inserting the cost parameters, the expected cost per cycle for *CCC* chart can be formulated as:

$$\begin{aligned} E(C_{cyc}) = {} & \frac{C_0}{\lambda} + C_1(-\tau + E + h(ARL2) + \delta_1 T_1 + \delta_2 T_2) + \frac{sY}{ARL1} \\ & + W + (a+b) \times \frac{(1/\lambda - \tau + E + h(ARL2) + \delta_1 T_1 + \delta_2 T_2)}{h} \end{aligned} \tag{9.2}$$

Note that since this is a renewal reward process, the expected cost per hour can be computed as the ratio of expected cost per cycle to the expected cycle time in hours $E(C) = E(C_{cyc}) / E(T)$. The expected cost per hour $E(C)$ for *CCC* chart, as a function of inspection interval and control limit, can be derived as follows:

$$C(l,h) = C_0 + \frac{S_1}{D} + \frac{a+b}{h} \times \frac{S_2}{D} \tag{9.3}$$

where the expected number of samples while in control, s, is

$$s = \frac{1}{e^{\lambda h} - 1} \tag{9.4}$$

and the expected time the process goes out-of-control between two samples, τ, is

$$\tau = \frac{1}{\lambda} - \frac{h}{e^{\lambda h} - 1} \tag{9.5}$$

Furthermore, it can be shown that

$$\begin{aligned} S_1 &= \lambda(C_1 - C_0)(-\tau + E + h(ARL2) + T_1 + T_2) \\ &- \lambda C_1[(1-\delta_1)T_1 + (1-\delta_2)T_2] + \frac{\lambda s Y}{ARL1} - \frac{\lambda C_0(1-\delta_1)sT_0}{ARL1} + \lambda W \end{aligned} \tag{9.6}$$

$$D = 1 + \frac{(1-\delta_1)sT_0}{ARL1} + \lambda(-\tau + E + h(ARL2) + T_1 + T_2)$$

and

$$S_2 = 1 + \lambda(-\tau + E + h(ARL2) + \delta_1 T_1 + \delta_2 T_2) \tag{9.7}$$

9.2.2. An optimization procedure

The algorithm by Lorenzen and Vance (1986) to find the most economical design is somewhat complicated. It consists of Newton's Method, golden section search, and Fibonacci search. This could be the main reason that limits the application of their method. Furthermore, they expand the exponential in τ, $\partial\tau/\partial h$, s and $\partial s/\partial h$, and ignore all terms containing powers of λ greater than 1. Hence, the solution may not be accurate despite its complexity.

In Chung (1991), a simplified procedure is developed to determine the optimum parameter values based on the LV model. The procedure, which has been further investigated and applied in a series of papers, e.g., Chung (1992, 1995), consists of solving an explicit formula for sampling interval h in terms of sample size n and control limit l. It yields lower loss costs than those of LV model. Furthermore, no terms are neglected for finding the optimum solution. Consequently, it was demonstrated that the procedure gives more accurate, more reliable and more feasible optimum values. Hence, it is more appropriate to use this simplified procedure in the economic design of *CCC* chart and will be described here.

For all $\lambda > 0$, $h > 0$, we can obtain

$$1 > \frac{1}{\lambda h} - \frac{1}{e^{\lambda h} - 1} > 0 \tag{9.8}$$

when the Taylor expansion $e^{x} = 1 + x + \frac{x^2}{2!} + \frac{x^3}{3!} + \cdots$ is used.

Generally, λh is a small value, so $(1/\lambda h - 0.5)$ is a good approximation to $1/(e^{\lambda h} - 1)$ because

$$\lim_{\lambda h \to 0}\left(\frac{1}{\lambda h} - \frac{1}{e^{\lambda h} - 1}\right) = \lim_{\lambda h \to 0} \frac{\lambda^2 h^2 / 2! + \cdots}{\lambda h(\lambda h + \lambda^2 h^2 / 2! + \cdots)} = 0.5 \tag{9.9}$$

For a given lower control limit l, replacing $1/(e^{\lambda h} - 1)$ with approximation $1/\lambda h - 0.5$, an explicit formula of h for *CCC* chart can be shown as:

$$h = \frac{-r_1 + \left(r_2^2 - 4 r_1 r_3\right)^{0.5}}{2 r_1} \tag{9.10}$$

where

$$r_1 = \lambda(ARL2 - 0.5)\left\{(C_1 - C_0)\left(1 - \frac{\lambda(1-\delta_1)T_0}{2ARL1}\right) + \frac{\lambda Y}{2ARL1}\right.$$
$$\left. + \lambda C_1\left[(1-\delta_1)T_1 + (1-\delta_2)T_2\right] - \frac{\lambda C_0(1-\delta_1)T_0}{2ARL1} - \lambda W\right\}$$
$$-(a+b)\left[\lambda(ARL2 - 0.5)\right]^2$$

$$r_2 = 2\lambda(ARL2 - 0.5)\{\frac{C_0(1-\delta_1)T_0 - Y}{ARL1}$$
$$-(a+b)(1 + \lambda E + \lambda\delta_1 T_1 + \lambda\delta_2 T_2)\}$$

and

$$r_3 = \frac{C_0(1-\delta_1)T_0 - Y}{ARL1}\left(1 - \frac{\lambda(1-\delta_1)T_0}{ARL1} + \lambda E + \lambda T_1 + \lambda T_2\right) + \frac{(1-\delta_1)T_0}{ARL1}$$
$$\{\lambda(C_1 - C_0)(E + T_1 + T_2) - \lambda C_1[(1-\delta_1)T_1 + (1-\delta_2)T_2] - \frac{\lambda Y}{2ARL1}$$
$$+ \frac{\lambda C_0(1-\delta_1)T_0}{2ARL1} + \lambda W\Big\} + \frac{(a+b)\lambda(ARL2 - 0.5)(1-\delta_1)T_0}{ARL1} - (a+b)$$
$$\cdot(1 + \lambda E + \lambda\delta_1 T_1 + \lambda\delta_2 T_2)\left(1 - \frac{\lambda(1-\delta_1)T_0}{2ARL1} + \lambda E + \lambda T_1 + \lambda T_2\right)$$

As the above formulas show, the loss function depends on $ARL1$, $ARL2$ and h which can be solved in terms of $ARL1$ and $ARL2$, which in turn are functions of control limit l.

The probability of *CCC* data getting larger than the lower control limit, that is, the probability of not sounding an alarm with one-side *CCC* chart is

$$P_{na} = \sum_{i=l}^{\infty}(1-p)^{i-1}p = (1-p)^{l-1} \tag{9.11}$$

Hence the chance for *CCC* data to be out of control limits, i.e., smaller than l, is $(1-P_{na})$. Therefore, only one out of $1/(1-P_{na})$ cumulative counts falls outside the lower control limit on the average. The average number of points plotted between two consecutive alarms is $E(P) = 1/(1-P_{na})$.

For a *CCC* chart, by using geometric distribution, it can be shown that

$$ARL = l + \frac{1-p}{p} + \frac{1}{p[1-(1-p)^{l-1}]} \tag{9.12}$$

When the process performs "in-control" state with the fraction nonconforming item p_0, the average number of samples taken before an alarm is

$$ARL1 = l + \frac{1-p_0}{p_0} + \frac{1}{p_0[1-(1-p_0)^{l-1}]} \tag{9.13}$$

When the fraction nonconforming item of process is shifted to p_1, the ARL is

$$ARL2 = l + \frac{1-p_1}{p_1} + \frac{1}{p_1[1-(1-p_1)^{l-1}]} \tag{9.14}$$

Based on the Chung's simplified procedure and the derived ARL, we may summarize the procedure of locating the optimal design parameters as follows:

(1) Obtain a value of l_0 which belongs to a reasonable searching range. (i.e. $l_a \le l_0 \le l_b$);

(2) Compute the $ARL1$ and $ARL2$ using (9.13) and (9.14) with this value of l;

(3) Compute h_0 from Eq. (9.10);

(4) Compute $C(l_0, h_0)$ from Eq. (9.3);

(5) Repeat steps 1-4 and find $C^* = \min_{l_a \le l_0 \le l_b} [C(l_0, h_0)] = C(l^*, h^*)$.

The values l^* and h^* which give the overall approximate minimum loss cost C^* constitute the optimal economic design for the CCC control chart.

9.2.3. Numerical example

Lorenzen and Vance (1986) presented an application of their economic model of a foundry operation that produces 84 casting per hour on one of its lines. This example will be used in the rest of this chapter, and we briefly describe the background, and also provide the basic input data here.

As explained in Lorenzen and Vance (1986), in castings, a standard is set to prevent high carbon-silicate content as it will result in low tensile strength. The sampling process costs \$4.22 per sample and takes about five minutes. The process stays in control for 50 hours. When the process is out of control, the system must be flushed and restarted, and this should take about 45 minutes with a repair crew cost of \$22.80 per hour. It is also assumed a downtime cost of \$21.34 per minute. It takes about five minutes to get the repair crew assembled and no time to search for the cause.

Example 9.1:
Here we assume that the case occurs in a high quality process and the *CCC* chart is used. Suppose that the historical data indicate that the process produces about 0.05% nonconforming items when in control and about 5% nonconforming items when out of control. The cost per hour while the process is in control is $C_0 = 100 \times 84 \times 0.0005 = \$4.2$. Similarly, C_1 is equal to \$420. Other specific input parameter values are $\lambda = 1/50$ hours, $E = T_0 = T_1 = 5/60$ hours, $T_2 = 45/60$ hours, $\delta_1 = 1$, $\delta_2 = 0$, $Y = W = \$977.40$, $a = 0$, $b = 4.22$.

From Eq. (9.3), the loss function approaches gives an optimal value of $C^* = 77.4763$ when the sampling interval is $h^* = 0.3015$ and the control limit is $l^* = 96.97$. The average run lengths are computed as $ARL1^* = 1.057 \times 10^6$ and $ARL2^* = 357.56$, respectively. The probability of false alarm when the process is in control is $\alpha^* = 0.0096$, and the probability of failing to detect the shift when the process is out of control is $\beta^* = 0.3812$.

Table 9.1 shows the economic designs of *CCC* charts with varied values of p_0 and p_1. The other parameters, such as $ARL1$ which represents average run length when process is in control and $ARL2$ which represents average run length when process is out of control, the probability of false alarms α and the probability of failing to detect the assignable cause β are also listed.

□

Table 9.1. The economic design of the *CCC* charts for different p_0 and p_1 values

p_0	p_1	l^*	h^*	ARL_1^*	ARL_2^*	C^*	α^*	β^*
0.0001	0.01	96.96	0.0333	1057200	357.56	413.86	0.0095	0.3812
0.0001	0.009	107.64	0.0321	952780	397.34	427.46	0.0106	0.3813
0.0001	0.008	121.00	0.0306	848410	447.07	443.21	0.0119	0.3814
0.0001	0.007	138.18	0.0292	744110	511.01	461.90	0.0136	0.3815
0.0001	0.006	161.07	0.0277	639860	596.26	484.60	0.0158	0.3816
0.0001	0.005	193.13	0.0261	535680	715.61	513.02	0.0190	0.3817
0.0001	0.004	241.20	0.0244	431560	894.63	550.10	0.0237	0.3818
0.0001	0.003	321.31	0.226	327530	1193.0	601.48	0.0315	0.3820
0.0001	0.002	481.50	0.0210	223630	1789.7	679.79	0.0469	0.3821
0.0001	0.001	962.08	0.0214	120020	3580	820.08	0.0916	0.3822
0.0002	0.001	960.01	0.0215	34605	3580	820.44	0.1745	0.3830
0.0003	0.001	958.08	0.0216	17644	3580	820.93	0.2496	0.3838
0.0004	0.001	956.78	0.0217	11323	3580	821.49	0.3177	0.3843
0.0005	0.001	956.3	0.0219	8220	3580	822.07	0.3798	0.3845
0.0006	0.001	956.5	0.0220	6440.9	3580	822.7	0.4364	0.3844
0.0007	0.001	957.5	0.0221	5311	3580	823.2	0.4882	0.3841
0.0008	0.001	958.9	0.0223	4542.5	3580	823.7	0.5355	0.3835
0.0009	0.001	960.9	0.0224	3991.2	3580	824.26	0.5786	0.3827

From Table 9.1, it can be noted that the optimal loss C^* is a decreasing function of p_1/p_0. For a given value of p_0, the optimal sampling interval h^* is increasing with value of p_1, and for a given value of p_1, h^* increases when the value of p_0 increases. The situation of optimal value of l^* is opposite to that of h^*. That is, for a given value of p_0, the optimal control limit increases with value of p_1, and for a given value of p_1, l^* increases when the value of p_0 increases. The above analysis is unsuitable for a very small shift, say, $p_1/p_0 < 2$. For instance, as last five lines in Table 9.1 show, l^* is increasing when the value of p_0 increases.

Both $ARL1$ and $ARL2$ are the increasing functions of the ratio of p_1 and p_0. When the ratio is relatively large, the probability of false alarm is small. However, when the ratio is very small, for example, smaller than 2, the probability of false alarm will increase dramatically. Especially, when the ratio approaches one, the probability of false alarm reaches 58%. Table 9.1 also shows that the probability of economic design failing to detect the assignable cause almost remains constant in the whole range of ratio.

9.2.4. A sensitivity analysis

There are many parameters involved in the loss function, and most of the cost parameters and the process parameters are difficult to estimate accurately in practice. The purpose of the sensitivity analysis is to determine the key drivers of cost and two *CCC* control chart decision variables. Based on the results from the sensitivity analysis, one may take more effort on estimating the key parameters accurately in practice, which will greatly reduce the difficulties of implementing economic design.

When applying Lorenzen and Vance model in the economic design of *CCC* chart, there are fifteen input parameters related to the loss function. The parameters, p_0 and p_1 are the process parameters which specify the in-control and out-of-control states of the process monitored. From numerical studies, we have noted that the there is almost no cost penalty associated with incorrectly specified p_0, but a relatively large cost penalty if p_1 is incorrectly estimated.

If we only consider other input parameters, by adopting the approach used in Costa (1993) and Bai and Lee (1998), sensitivity analysis of the proposed model with regard to these input parameters can be easily made. Here we only present some general results. Detailed tables, which can be found in Tang (2000), are omitted.

The parameter λ, which is the average occurrence rate of the assignable cause, has a significant impact on the optimum sampling interval and the cost. A smaller λ implies a larger optimal sampling interval and smaller cost. Time to sample and chart one item, *E*, expected time to discover the assignable cause, T_1, and expected time to repair the process, T_2, have very slight effects on the optimal values.

In summary, the optimum control limit l^* is relatively robust to the input parameters, except for p_0 and p_1. Among all other factors, C_1, C_0 and λ are the most sensitive factors which affect the optimal cost and optimal sampling interval for *CCC* chart. This may help the user of economic design procedure to carefully allocate the resource in obtaining the best estimate of the input parameters.

Although the optimal economic design method might produce the optimal sampling interval for *CCC* control chart, the specified value of h may not be easily achieved in practice. Some processes can only be sampled at fixed intervals. Particularly in those processes with 100% inspection, the sampling intervals are strictly constrained by the production rate. Therefore, it is much more convenient to sample at the proposed time interval.

9.3. Economic-statistical design of *CCC* chart

In the previous sections, the design parameters are determined purely according to the economic concern of the process being monitored. Although economic design of *CCC* chart has provided a method to optimize the risk of false alarms and that of failing to detect the process shift from an economic point of view, statistical properties have been sacrificed to a great extent. For example, the risk of failing to detect the shift is almost 38% even when shift is fairly large. That is, more than every third shift cannot be detected, which implies that economically designed *CCC* chart is not very sensitive in detecting shifts.

The purpose of this section is to develop and study an economic-statistical design model for *CCC* control charts, and also to give an optimal search procedure to find the design parameters. The properties of economic-statistical designed *CCC* control charts are then presented together with some comparisons using other existing design methods. This may provide an alternative way to design control charts for high quality processes. The charting and optimization procedure can be readily implemented in a computed-aided manufacturing environment.

9.3.1. Economic-statistical model for *CCC* chart

The economic statistical design of a control chart is defined as a design in which the loss cost function is minimized subject to some statistical constraints. In this chapter, maximum values of Type I error and Type II error are placed to control the statistical behavior of the *CCC* chart.

Let $C(h, ARL1, ARL2)$ be the loss function of the economic model presented in Section 3.2. The model for economic-statistical design for CCC chart can be formulated as:

$$\text{Minimize } C(h, ARL1, ARL2) \tag{9.15}$$

$$\text{Subject to: } \alpha \le \alpha_u \text{ and } \beta \le \beta_u$$

where α and β represent, respectively, Type I error and Type II error at the expected shift level provided by the economic-statistically designed chart.

As we have noted, the economic-statistical design involves the selection of the upper bounds of Type I error and Type II error. Suppose the fraction nonconforming of a process is p_0, and the cumulative count of conforming items follows a geometric distribution with mean $1/p_0$. The probability for CCC to be not smaller than the control limit l, that is, the probability of not sounding an alarm is

$$\sum_{i=l}^{\infty}(1-p_0)^{i-1}p_0 = (1-p_0)^{l-1} \tag{9.16}$$

The low-sided CCC chart is of more practical interest; thus we will mainly focus on the lower control limit which is given by

$$l = LCL = \ln(1-\alpha)/\ln(1-p) \tag{9.17}$$

where p is the process fraction nonconforming level and α is the preset alarm rate.

When the process is in control, the probability of CCC chart having the false alarm rate is

$$\alpha = 1-(1-p_0)^{l-1} \tag{9.18}$$

When the process shift to another level p_1, the Type II error β is

$$\beta = (1-p_1)^{l-1} \tag{9.19}$$

Note that the economic-statistical model of *CCC* charts satisfies the constraints:

$$1-(1-p_0)^{l-1} \le \alpha_u \tag{9.20}$$

and

$$(1-p_1)^{l-1} \le \beta_u \tag{9.21}$$

Because of this, we have the following requirements:

$$\frac{\ln(1-\alpha_u)}{\ln \beta_u} \ge \frac{\ln(1-p_0)}{\ln(1-p_1)} \tag{9.22}$$

It should be noted that the above inequity may not be met for given values of alarm rate and process fraction nonconforming. For example, in the case that $p_0 = 0.0001$, $p_1 = 0.001$, $\alpha_u = 0.1$ and $\beta_u = 0.3$, we have

$$\frac{\ln(1-\alpha_u)}{\ln \beta_u} = 0.0875 \text{ and } \frac{\ln(1-p_0)}{\ln(1-p_1)} = 0.100$$

As a result, one cannot find the optimum design parameters with such fixed bounds of Type I error and Type II error. In such a situation, we may have to relax the bounds of Type I error and Type II error.

The final objective of the optimization procedure is to find the optimal control limit. In the process of obtaining approximate optimum values l^* and h^*, l can be considered as a discrete variable. The complete procedure is given as follows.

(1) Select the bounds of Type I error and Type II error, and use Eq. (9.22) to check if there is feasible solutions. If it can be satisfied, go to Step 2; otherwise, reconsider the bounds or stop the procedure;

(2) Obtain a value of l_0 which belongs to a reasonable searching range ($l_a \le l_0 \le l_b$);

(3) Calculate the Type I error and Type II error from (9.18) and (9.19). If any of (9.20) and (9.21) cannot be satisfied, i.e., there is no feasible solution with this special l_0, go back to step 2; otherwise, go to Step 4;

(4) Compute the $ARL1_0$ and $ARL2_0$ using (9.13) and (9.14) with this value of l_0;

(5) Compute h_0 from Eq. (9.10);

(6) Compute $C(h_0, ARL1_0, ARL2_0)$;

(7) Repeat the steps 2-6, and find

$$C^* = \min_{l_a \le l_0 \le l_b} [C(l_0, ARL1_0, ARL2_0)] = C(h^*, ARL_1^*, ARL_2^*);$$

9.3.2. A numerical example

The specific input parameter values earlier cannot be used to serve an illustrative purpose directly since the notations are slightly different. However, McWilliams (1996) presented some relationships between different models and notations. In this case, the following parameters can be obtained: $\lambda = 0.01$, $E = 0$, $T_1 + T_2 = 0.3$, $\delta_1 = 0$, $\delta_2 = 0$, $C_0 = 0$, $C_1 = 100$, $Y = 25$, $W = 75$, $a = 0$, $b = 0.5$. These will be used as input parameter here.

Example 9.2:

Suppose a process starts from an "in control" state with the fraction nonconforming $p_0 = 0.0005$. When an assignable cause occurs, the fraction nonconforming will change to $p_1 = 0.005$. If the statistical constraints are selected as $\alpha_u = 0.15$ and $\beta_u = 0.25$, then the results can be easily obtained using the above optimization procedure. The loss function approaches its optimal value $C^* = 35.378$; the sample interval is $h^* = 0.0323$ and the lower control limit $l^* = 277.565$.

The ARL at the acceptable level is $ARL1^* = 17759$ and the ARL at the reject level will be $ARL2^* = 743$. Furthermore, the probability of false alarms when the process is in control is $\alpha^* = 0.1292$, while the probability of failing to detecting the shift when the process is out of control will be $\beta^* = 0.25$.

□

A comparison

Here we present a comparison between economic statistical design of *CCC* chart and pure economic design. Some results are shown in Table 9.2. The economic statistical design has larger control limits, smaller sampling interval, and higher costs.

The β for the economic statistical design could be far better than the corresponding values for the economic design. In other words, economic statistically designed *CCC* charts are more powerful in detecting shifts. One negative fact we may notice is that the Type I errors and the costs for the economic statistical designs listed in Table 9.2 are relatively higher. However, in view of the substantial improvement in the power of detecting shifts, we may conclude that this shortcoming can be justified.

When applying the economic design methodology, it is useful to conduct sensitivity analyses regarding the effects of the cost and operating parameters on the optimal design parameters. In carrying out a sensitivity analysis, statistical design of experiments can be used. Although p_0 is also very important to the economically optimum design, it is more useful to study the performance of economic design when p_0/p_1 varies. Tang *et al.* (2000) presented some examples to which interested readers can refer for more details.

Table 9.2. A comparison with economic design

	Economic Statistical Design					Economic Design				
No	C^*	h^*	l^*	α^*	β^*	C^*	h^*	l^*	α^*	β^*
1	28.02	0.0397	230.11	0.0227	0.1	25.53	0.0434	102	0.01	0.36
2	29.38	0.038	255.69	0.0251	0.1	26.6	0.0415	108	0.0106	0.38
3	30.96	0.0362	287.67	0.0283	0.1	28.22	0.0395	121	0.0119	0.38
4	32.86	0.0344	328.79	0.0322	0.1	29.96	0.0374	138	0.0136	0.38
5	35.16	0.0324	383.61	0.0375	0.1	32.08	0.0352	161	0.0159	0.38
6	38.07	0.0302	460.36	0.0449	0.1	34.77	0.0327	193	0.019	0.38
7	41.91	0.0279	575.49	0.0558	0.1	38.32	0.0301	241	0.0237	0.38
8	47.31	0.0254	767.38	0.0738	0.1	43.34	0.0272	322	0.0316	0.38
9	53.79	0.0231	948.61	0.0904	0.15	51.29	0.0240	482	0.0470	0.38
10	67.96	0.0204	1386.6	0.1294	0.25	67.03	0.0206	963	0.0917	0.38

9.4. Some further extensions

The economic design of *CCC* charts is extended in the following two directions:

i) relaxing of some assumptions, and

ii) using a generalization of *CCC* chart.

Both economic and economic-statistical designs of *CCC* charts require a distribution of process failure mechanism to be specified. In order to simplify the economic model development, a Poisson process, which indicates a constant failure rate, is assumed in the previous study. However, this assumption is not always appropriate in practice, especially for processes in which machine and tool wear occurs over time. Processes with an increasing hazard function are quite common in modern industries. Therefore, it is of interest to investigate the performance of economically designed *CCC* charts under a Weibull failure mechanism.

In process monitoring, detecting both process deterioration and process improvement is equally important. Although the one-sided *CCC* chart is effective in detecting process deterioration, the economically designed two-sided *CCC* chart is of practical interest. The major reason is that it can also detect process improvement with both economic and statistical benefits. In addition, the *CCC* chart is just one special case of *CCC-r* charts.

9.4.1. Weibull failure mechanism

The purpose of this section is to study the economic design of *CCC* charts under the assumption of Weibull distribution. Another aim is to develop and study the economic models for the general type of *CCC* charts. The Weibull process failure mechanism is examined first. Thereafter is the economic design of the one-sided *CCC-r* chart. The economic design of the two-sided *CCC* chart is then presented. Comparative studies among these economic designs are also given.

The advantage of the Weibull distribution is that it can take on a variety of shapes and mimic the behavior of other distributions. Compared to the exponential distribution, Weibull distribution is more flexible in that it includes models with increasing or decreasing hazard rate. In the literature, the research on process failure mechanism assumption for the economic designs of *CCC* charts has not been considered. However, it is useful to examine the robustness of the process failure mechanism assumption is on *CCC* charts. Knowledge of the robustness is useful to practitioners when they implement such designs.

Banerjee and Rahim (1988) compared the process failure mechanisms, which followed a Weibull distribution and exponential distribution under a dynamic sampling interval policy. They concluded that hourly cost could be reduced according to the parameters in Weibull distribution. However, it is very difficult to implement control charts in practice when the sampling interval is determined dynamically. For simplicity, it is assumed that the sampling interval is held constant in this study. Moreover, this is the most likely scenario in the actual implementation. Zhang and Berardi (1997) and Costa and Rahim (2000) are some recent papers using the Weibull failure model in economic design of control chart.

The economic design of a *CCC* control chart can be easily modified to incorporate a Weibull distribution for the process failure mechanism. All the other notations and assumptions remain the same as those in Section 9.2. In the economic model, replace $1/\lambda$ with μ, where μ is the mean of the Weibull distribution. Following the earlier notations, the expected cost per hour of the *CCC* chart under Weibull distribution can be obtained as:

$$E(C) = \frac{\frac{C_0}{\mu} + C_1\left(-\tau + E + h(ARL2) + \delta_1 T_1 + \delta_2 T_2\right) + \frac{sY}{ARL1} + W}{\frac{1}{\mu} + \frac{(1-\delta_1)sT_0}{ARL1} - \tau + E + hARL2 + T_1 + T_2} + (a+b) \times \frac{\left(1/\mu - \tau + E + h(ARL2) + \delta_1 T_1 + \delta_2 T_2\right)}{h\left(\frac{1}{\mu} + \frac{(1-\delta_1)sT_0}{ARL1} - \tau + E + hARL2 + T_1 + T_2\right)} \tag{9.23}$$

The other components in the model that need to be changed are only s and τ. They are calculated with more complex formulas when compared to the exponential distribution process failure mechanism.

Let T_A, which denotes the random variable that characterizes the time until an assignable cause occurs, follow a Weibull distribution with the scale parameter γ and the shape parameter ν. Then we have

$$s = \sum_{j=0}^{\infty} jP\{jh < T_A \leq (j+1)h\} = \sum_{j=0}^{\infty} \exp[-(\gamma jh)^{v}] \tag{9.24}$$

while $\tau = \mu - hs$.

Example 9.3:
In Table 9.1, a total of 18 examples of the optimal *CCC* chart design parameters for various values of p_0 and p_1 were shown. Ten of these examples are used here to illustrate the effects of using the more general Weibull distribution on the optimal plan parameters and on the resulting expected hourly cost.

The mean time to the occurrence of the assignable cause is maintained at $\mu = 50$ hours, but the shape parameters vary from 0.5 and 2.0. Among them, 0.5 represents decreasing hazard rates and 2.0 represents increasing hazard rates. Table 9.3 presents an analysis on the sensitivity of the shape parameter for economic design. The control limit, sampling interval, Type I and II errors, and cost are described in the table.

□

The results of this and other numerical studies give strong evidence that the economic design of *CCC* charts is insensitive to the assumed process mechanism since there is no substantial change in the optimal design parameters and costs. Therefore, the requirement to correctly specify the process failure mechanism could be not so important for *CCC* charts in practice.

Table 9.3. Sensitivity of the shape parameter v for the *CCC* economic design model

$v = 0.5$						
p_0	p_1	l^*	h^*	α^*	β^*	C^*
0.0001	0.01	96.9552	0.0333	0.0096	0.3812	413.96
0.0001	0.009	107.6426	0.032	0.0106	0.3813	427.47
0.0001	0.008	121.0012	0.0306	0.0119	0.3814	443.21
0.0001	0.007	138.1754	0.0292	0.0136	0.3815	461.90
0.0001	0.006	161.0725	0.0277	0.0159	0.3816	484.60
0.0001	0.005	193.1254	0.0261	0.019	0.3817	513.02
0.0001	0.004	241.1988	0.0244	0.0237	0.3819	550.10
0.0001	0.003	321.3083	0.0226	0.0315	0.382	601.48
0.0001	0.002	481.4987	0.021	0.0469	0.3821	679.79
0.0001	0.001	962.0827	0.0214	0.0916	0.3823	820.08

$v = 2$						
p_0	p_1	l^*	h^*	α^*	β^*	C^*
0.0001	0.01	96.9552	0.0333	0.0096	0.3812	413.96
0.0001	0.009	107.6426	0.032	0.0106	0.3813	427.47
0.0001	0.008	121.0012	0.0306	0.0119	0.3814	443.21
0.0001	0.007	138.1754	0.0292	0.0136	0.3815	461.90
0.0001	0.006	161.0725	0.0277	0.0159	0.3816	484.60
0.0001	0.005	193.1254	0.0261	0.019	0.3817	513.02
0.0001	0.004	241.1988	0.0244	0.0237	0.3819	550.10
0.0001	0.003	321.3083	0.0226	0.0315	0.382	601.48
0.0001	0.002	481.4987	0.021	0.0469	0.3821	679.79
0.0001	0.001	962.0827	0.0214	0.0916	0.3823	820.08

9.4.2. Economic design of one-sided *CCC-r* chart

If we count the number of items inspected till observing r nonconforming items and monitor the process by this count, we have the *CCC-r* chart. This chart was discussed in Section 5.2. The economic concept can also be applied to the general *CCC-r* charts, so that comparison study can be carried out from an economic point of view. Among others, this type of study will be very useful in the selection of parameter r.

The probability for the count *CCC-r* to be out of the lower control limit is $(1-P_{na})$. Thus, on the average, only one out of $1/(1-P_{na})$ falls outside of LCL_r. That is, the average number of points plotted on the *CCC-r* chart between two consecutive alarm signals is

$$E(P)=\frac{1}{1-P_{na}} \tag{9.25}$$

It can be shown that when the process is assumed to start from an "in control" state with a proportion of nonconforming item p_0, the ARL in control can be calculated as:

$$ARL1_r=\frac{1}{1-P_{na}}\times\frac{r}{p_0}+\sum_{i=LCL_r}^{\infty}\left(\frac{i}{P_{na}}\right)\binom{i-1}{r-1}p_0{}^r(1-p_0)^{i-r} \tag{9.26}$$

When process deterioration results in the fraction of nonconforming item to be p_1, the ARL out of control will be:

$$ARL2_r=\frac{1}{1-P_{na}}\times\frac{r}{p_1}+\sum_{i=LCL_r}^{\infty}\left(\frac{i}{P_{na}}\right)\binom{i-1}{r-1}p_1{}^r(1-p_1)^{i-r} \tag{9.27}$$

As shown in Xie *et al.* (1999), the control limits of *CCC-r* chart increase very quickly when the process fraction nonconforming p approaches zero. When p is small enough, the value of the central line, which indicates the average items inspected until a point is plotted, will become very large. Thus, it is not appropriate to have a large r value. Xie *et al.* (1999) suggested using a r value from 2 to 5.

Taking $r = 2$ as an example, we have

$$ARL_2 = \frac{2 - 4p + 2p^2 + 2lp - 3lp^2 + l^2 p^2}{lp - 2p^2 + lp^2} + \frac{2}{p[1 - (1-p)^{l-2}(1 - 2p + lp)]} \quad (9.28)$$

Example 9.4:

Consider the economic design of a *CCC* chart with $r = 2$. The cost information is assumed to be obtained as in Lorenzen and Vance (1986): $\lambda = 1/50$ hours, $E = T_0 = T_1 = 5/60$ hours, $T_2 = 45/60$ hours, $C_0 = \$114.24$, $C_0 = \$114.24$, $C_1 = \$949.20$, $\delta_1 = 1$, $\delta_2 = 0$, $Y = W = \$977.40$, $a = 0$, and $b = \$4.22$.

Table 9.4 shows the optimum setting of design parameters to minimize cost without any statistical constraints for the different values of p_0 and p_1. It can be seen that the properties of the economically designed *CCC*-2 are similar to those of the economically designed *CCC* chart.

□

Note that in our investigation, we have fixed the value of *r* here. However, it could be of interest to let *r* be a design parameter as well. A recent study on the design of *CCC-r* chart is Ohta *et al.* (2001). On the other hand, the selection of *r* could depend on several practical issues discussed in Section 6.1.

Table 9.4. The economic design of the *CCC*-2 charts for some different p_0 and p_1 values

p_0	p_1	l^*	h^*	ARL_1^*	ARL_2^*	C^*
0.0001	0.01	240.43	0.0268	71203000	658	499.63
0.0001	0.009	267.07	0.0259	57741000	731	516.38
0.0001	0.008	300.37	0.025	45697000	822	535.77
0.0001	0.007	343.18	0.024	35068000	940	558.59
0.0001	0.006	400.26	0.0231	25850000	1097	586.00
0.0001	0.005	480.17	0.0221	18041000	1316	619.81
0.0001	0.004	600.03	0.0212	11638000	1646	663.01
0.0001	0.003	799.8	0.0206	6638000	2195	720.86
0.0001	0.002	1199.3	0.021	3038300	3292	803.34
0.0001	0.001	2397.9	0.0344	836270	6585	922.03
0.0002	0.001	2396.9	0.0344	129900	6585	922.06
0.0003	0.001	2395.5	0.0345	48770	6585	922.13
0.0004	0.001	2394.3	0.0347	26288	6585	922.21
0.0005	0.001	2393.7	0.0348	17205	6585	922.31
0.0006	0.001	2393.7	0.035	12672	6585	922.42
0.0007	0.001	2394.3	0.0351	10080	6585	922.52
0.0008	0.001	2395.4	0.0352	8453	6585	922.61
0.0009	0.001	2396.7	0.0354	7360	6585	922.70

9.4.3. Economic design of two-sided *CCC* Chart

Until now, we have considered the economic design procedures applied to the low-sided *CCC* charts. Because of the simplicity and practicality, corrective actions will only be taken when the conforming count between two nonconforming items is below the lower control limit. A lower control limit in *CCC* chart detects process deterioration. However, the detection of process improvement is equally important although it is usually more difficult. The two-sided *CCC* chart can be useful in detecting both process improvement and process deterioration at the same time and will be briefly discussed here.

Again, as before, we assume that one item is checked every h hours of production. The number of consecutive conforming items is counted until a nonconforming one is detected. If the count is outside the lower control limit and the upper control limit, assignable caused should be searched for. Recall that at a preset false alarm α, the lower control limit l and the upper control limit u are

$$l = \ln(1-\alpha/2)/\ln(1-p) \tag{9.28}$$

and

$$u = \ln(\alpha/2)/\ln(1-p) \tag{9.29}$$

From the above, we see that, for a fixed lower control limit l, the upper control limit is related to the lower control limit as

$$u = \frac{\ln[1-(1-p)^l]}{\ln(1-p)} \tag{9.30}$$

This can be inserted to the loss function as the exact probability limits for the economically designated *CCC* chart. Although the new loss function appears not different from the one in the one-sided *CCC* chart, the introduction of the upper control limit will greatly change the *ARL* performance of the *CCC* chart.

The overall optimisation model is as before. The economic and economic-statistical designs of two-sided CCC charts are illustrated with two examples in the following. The first one considers the process deterioration. The second example illustrates the use of an economically designed two-sided *CCC* chart in the detection of process improvement.

Example 9.5:

In this example, assume process shifts occur according to an exponential distribution with the frequency of one every half hundred hours of operations. Hence $\lambda = 0.02$. Other cost and process parameters are same as those in Lorenzen and Vance (1986), i.e., $E = T_0 = T_1 = 5/60$ hours, $T_2 = 45/60$ hours, $C_0 = \$114.24$, $C_1 = \$949.20$, $\delta_1 = 1$, $\delta_2 = 0$, $Y = W = \$977.40$, $a = 0$, and $b = \$4.22$.

In this example, a case of process deteriotation is investigated. The process starts from an "in-control" state with the fraction nonconforming $p_0 = 0.0001$ or $100\,ppm$. The fraction nonconforming shifts to $p_1 = 0.01$ when an assignable cause occurs. After using the optimization procedure, we find that the loss function approaches its optimal value $C^* = 413.98$ when the sampling interval $h^* = 0.0333$, the lower control limit $l^* = 96.93$, and the upper control limit $u^* = 465000$.

The average run length can be computed to be $ARL1^* = 5.33 \times 10^5$ and $ARL2^* = 357.56$, respectively. The probability of false alarm when the process is in control is $\alpha^* = 0.0191$, and the probability of failing to detect the shift when the process is out of control is $\beta^* = 0.3813$.

□

Example 9.6:
In this example, we assume that process starts from an "in-control" state with the fraction nonconforming $p_0 = 0.01$, and the fraction nonconforming shifts to $p_1 = 0.001$ when an assignable cause occurs. This is the case when the process is improved, which is an emphasis in this book. By employing the two-sided model, the loss function approaches its optimal value $C^* =$ 647.6823 when the sample interval $h^* = 0.0248$, the lower control limit $l^* = 5.3256$, and the upper control limit $u^* = 314.1446$.

The ARL at the acceptable level is $ARL1^* =$ 1175.3 and the ARL at the reject level is $ARL2^* =$ 1361.3. Furthermore, the probability of false alarms when process is in control is $\alpha^* = 0.0851$, while the probability of failing to detect the shift when process is out of control is $\beta^* = 0.2654$.

□

9.4.4. Some other related topics

Economic design is a difficult but interesting issue in both research and applications. The cost is of ultimate concern to management when implementing quality procedures. On the other hand, many economic design procedures are not easy to follow and numerical computation is needed. Furthermore, the understanding of alarm rate and sensitivity is essential when interpreting the results. Another issue that is worth exploring is the effect of estimation (Lo, 1994).

Other than extending the economic design models by extending the current analytical research, it would also be of great interest to practitioners if simple procedures or numerical algorithms are developed. These procedures should have good statistical properties in the sense that the control chart design will yield near-optimal results.

Most of the results on economic design deal with univariate process characteristic. When multivariate characteristics are available, similar studies can be carried out although the cost functions will be more complicated. This is an interesting problem as when multiple characteristics are considered, the relative importance can be incorporated into the cost function (Jolayemi, 2000).

Economic design procedure can be developed for *EWMA-CCC* or *CUSUM-CCC* chart in a similar manner. The cost model discussed earlier can be adopted and modified to determine the economic design of such charts. Similar studies on *CQC* chart is also possible. In fact, it could be useful for continuous production process as the cost is usually a concern.

CHAPTER 10

Monitoring and Adjustment of Trended Processes

Although in this book we have focused on the statistical process control when quality characteristics are in form of attribute data, for effective analysis and quality improvement, accurate measurements can be very useful. In fact, the traditional X-bar and R-chart widely used in industry are based on variable quality characteristics. Standard techniques such as process capability analysis and variability reduction through design of experiments can be found in various texts and hence will not be discussed here. On the other hand, statistical process control techniques, especially the use of control chart for process monitoring, are usually based on the assumption that the process is stable with no significant trend.

In this section we briefly describe a common case when process is changing in the form of a predictable trend. Instead of carrying out complex analysis and advanced adjustment schemes such as advocated in engineering process control, we analyze the effect of the use of Shewhart chart for this type of processes. Although it is not advisable to use Shewhart chart for trended process, we argue that when the trend is slow and the adjustment is frequent, the process can be considered to be relatively stable. This leads to the question of quantifying the slow trend and frequent adjustments. In fact, a more interesting question on the determination of adjustment time will also be discussed.

10.1. Adjustment for process improvement

It is a well-known fact that processes usually deteriorate over time. This can be caused by what is considered to be not-removable causes such as tool wear or power consumption. For this type of processes, regular adjustment or replacement through the change of a tool, refill of power supply, or activation of a feedback controller, is a standard implementation practice, as illustrated in Figure 10.1. It is also well-known that regular maintenance will improve the long-term performance of a process. Although such a process is not rare in automated industry, the study of statistical monitoring for this type of processes is very limited. Little research has been reported on the use of control chart other than acceptance control chart first proposed in Freund (1957).

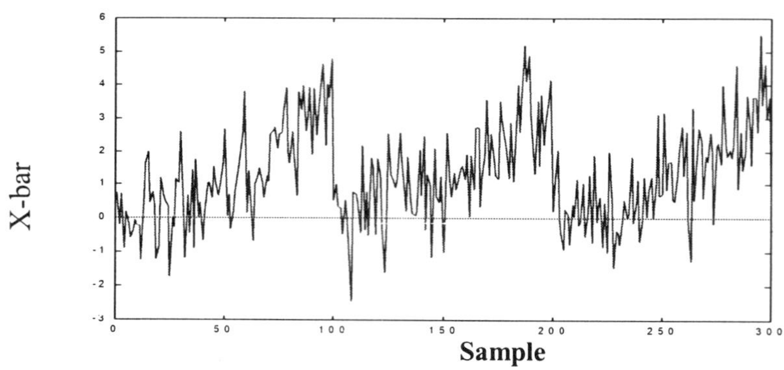

Figure 10.1. A typical trended and regularly adjusted process.

Control charts have been studied as a separate discipline, different from engineering control and adjustments. The design parameters of a control chart, i.e. sample size, sampling interval and control limit coefficients, are determined by statistical properties such as the power of statistical test, average run length (Woodall, 1985), and production cost over a non-adjusted span of production (Duncan, 1956). There is, however, an important issue for trended process concerning how and how often the process should be adjusted to maintain a certain quality level.

In industrial practices, many processes have to undergo regular engineering control to prevent violation of engineering specifications because of the intrinsic trend as mentioned before. In such cases, engineering control is not related to stoppage of a process and search of assignable causes, but could be part of operating procedure. The detailed form of engineering controllers may vary. We will not focus our discussion on the integration of SPC (statistical process control) and EPC (engineering process control) here and interested readers are referred to the work by Box and Luceno (1997); only a related case will be mentioned.

A simple approach is the dead-band adjustment scheme of Box and Luceno (1997) where the concept of average adjustment interval (*AAI*) is also introduced. For processes under dead-band adjustment, the concept of *AAI* in the SPC context is somewhat equivalent to that of *ARL* of control chart. The computation of *AAI* is also similar to that of *ARL* in the form of linear integral equation:

$$AAI_{tm} = m + \int_{|\hat{e}_{tm+m}|<L} AAI_{tm+m} \cdot \varphi(u)du \qquad (10.1)$$

where L is the boundary value, m is the sampling interval, $\hat{e}_{tm}$ is the predicted error at time point tm, and $\varphi(u)$ represents the probability density function of random disturbance u.

The dead-band approach is similar to SPC, for which the adjustment is carried out when the process values exceed a certain range. On the other hand, the SPC approach is easier. The effect, such as replacement of tools or power source, is considered to be fixed as it restores the process to its initial state. This is common in practice, but the question is how to deal with the case from an SPC point of view, i.e. whether Shewhart chart can be used and if so, how the properties can be interpreted.

10.2. Interpretation of traditional Shewhart chart

As discussed, when the process mean is drifting, the traditional Shewhart control chart for individual observations may not be appropriate. In this case, a trend-adjusted chart can be implemented (Figure 10.2). However, English (1994) indicated that using fixed control limits could be effective in many cases. Also, if there exists preventive engineering controls that keep the process mean relatively stable within a small acceptable range, it could be possible that the monitoring Shewhart chart would be relatively insensitive to the fluctuations of process mean within that range, and hence the Shewhart chart could still be applicable.

If a Shewhart chart is used when there is a slow trend, the false alarm probability will not be the same as that for a stable process which, with 3-sigma limits, should be about 0.27%. What is of interest is to study the actual properties when Shewhart chart is used and see how much it differs from the nominal value. When the difference is small, the use of the Shewhart chart is justified; if the difference is large, the Shewhart chart should not be used, or the process should be brought to a behavior of slow trend. This section aims to suggest ways in which the traditional Shewhart control chart can be interpreted.

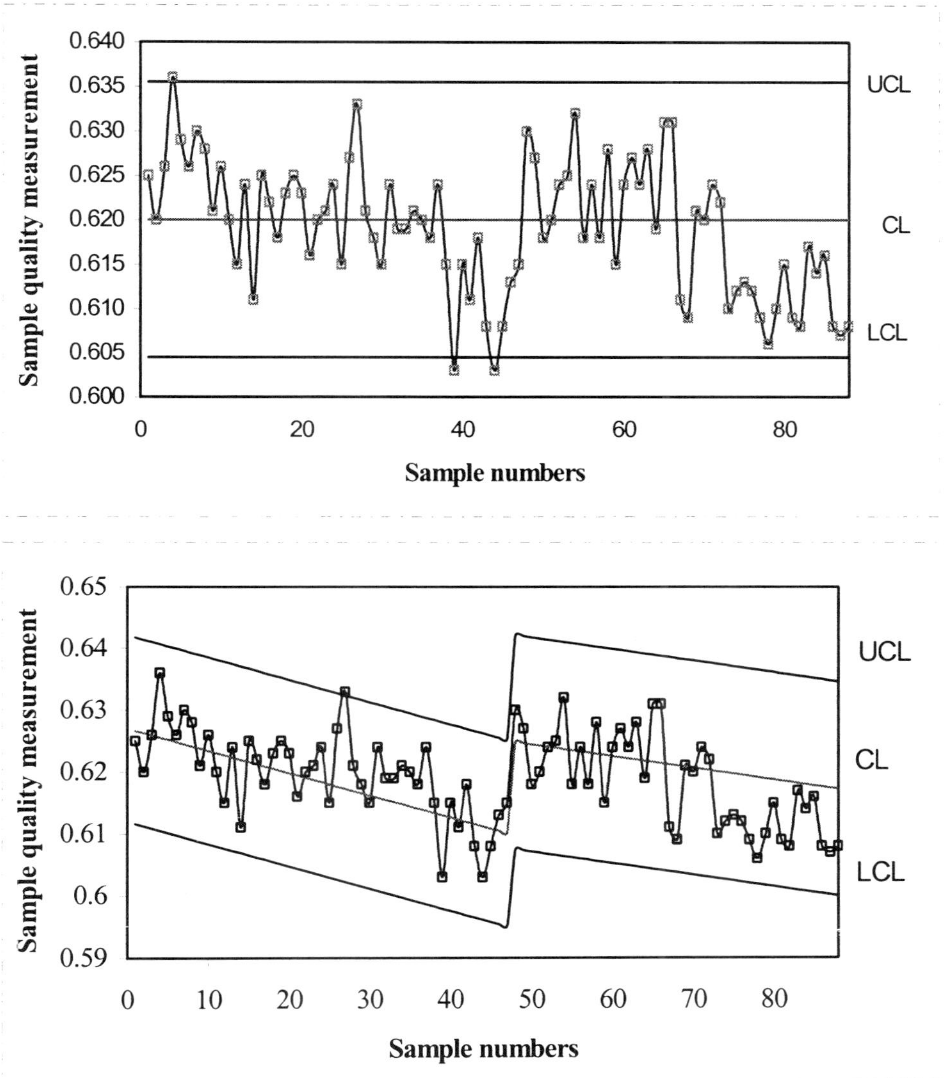

Figure 10.2. Monitoring of trended and adjusted process.

10.2.1. Some general results

Some general results are shown here without specifying a specific trend model. We assume that the trend is slow and adjustment relatively regular, so that the Shewhart chart is not obviously inappropriate. We investigate the false alarm rate and average run length when a Shewhart chart is used. The analysis can further be used to answer the question as to how slow is a slow trend and how frequent is a frequent adjustment. In addition, it can provide some help in determining the adjustment interval.

Alarm rate with Shewhart chart

Here the in-control state is defined as the state such that the process is not subject to assignable causes other than the trend that is incorporated in the regular adjustment scheme. The false alarm probability is an important property of any control chart; an alarm is triggered when a point is out of the control limits. When there are no assignable causes other than the nominal trend, which is considered acceptable either because it is not removable or because it is tolerated from a practical point of view, the alarm is considered to be a false alarm. The probability of a false alarm is the probability that a value is out of the control limits when the process behaves according to such "normal" or "in-control" behavior.

Xie *et al.* (2001a) investigated this case and introduced the concept of average alarm rate (*AAR*). The *AAR* is used because for processes with a deterministic trend, the probability that a value is out of fixed control limits changes with the sample number. This probability can be called Point Alarm Rates. The average of the alarm rate for each cycle, defined in the duration from the beginning of operation to the adjustment time, is the average alarm rate.

Assume that we have individual measurements, which is common in modern and automated manufacturing environments. For a given sample point *i,* the point alarm rate is given by

$$\alpha_i = P\{y_i > UCL\} + P\{y_i < LCL\} \tag{10.2}$$

where y_i is the sample measurement value and *UCL* and *LCL* are the Upper Control Limit and Lower Control Limit, respectively. This alarm rate depends on the sample number *i*. By taking the average over a cycle of length *I*, we can compute an *AAR* as

$$AAR(I) = \sum_{j=1}^{I} \alpha_j / I \tag{10.3}$$

Note that the average alarm rate is not the overall false alarm rate since it relates to a process that is not adjusted during *I* points. On the other hand, when the trend is slow or adjustment is frequent, this will be a good approximation of the average alarm rate.

Assume that a cycle begins when a point falls outside the control limits. The probability that the *j*:th point is the first that falls outside the control limits is

$$\begin{aligned} P_j = P\{[y_j > UCL \text{ or } y_j < LCL] \text{ and} \\ [UCL > y_i > LCL \text{ for } i = 1,2,\ldots, j-1]\} \end{aligned} \tag{10.4}$$

Here P_j is used to represent the probability of first issuing an alarm at sample point j.

The probability that point *j* is the first point that falls outside the control limits can be computed recursively as (Xie *et al.*, 2001a):

$$P_j = \alpha_j \cdot \frac{(1-\alpha_{j-1})}{\alpha_{j-1}} P_{j-1};\, j = 2,3,\ldots. \tag{10.5}$$

with the initial value given by

$$P_1 = \alpha_1 = P\{y_1 > UCL\} + P\{y_1 < LCL\}$$

Analytically, by using the recursive formula a number of times, we have the probability that point j is the first point at which an alarm is raised:

$$P_j = \alpha_j \cdot \frac{(1-\alpha_{j-1})}{\alpha_{j-1}} P_{j-1} = \alpha_j \cdot \prod_{i=1}^{j-1}(1-\alpha_i);\quad j = 2,3,.... \tag{10.6}$$

Again, no specific process trend model is assumed here. On the other hand, to compute the alarm probability, a specific model such as a linear one is needed as will be discussed later.

Average run length with Shewhart chart

The average run length, which is another important chart property, is defined as the expected number of points till an alarm signal. When the process has no trend and when 3-sigma limits are used, the in-control average run length is about 370. When the same limits are used for trended process, the average time to a false alarm could be shorter because of the increased alarm rate. However, when the trend is slow and the adjustments are frequent, it is expected that the average run length should not be too different from the standard value of 3-sigma limit, 370. When the difference is significant, this is an indication that other methods should be used.

Based on the probability of an alarm at sample number j, the expected length of a run without any point falling outside the control limits for a process that is adjusted only at fixed time interval I can be determined:

$$E[N] = \sum_{j=1}^{\infty} jP_j \tag{10.7}$$

However, when a control chart is used and when a point is outside the control limits, the common procedure is to check for assignable causes and restore the system to its original state immediately with a corrective adjustment. For a trended process, even if the process is not out-of-control, we might also regularly adjust the process and this could be interpreted as preventive adjustments. In this case, the average run length will be highly reduced, especially when the process is in control.

When the process is adjusted at a frequency of I sample points since the last adjustment, or at the time an out-of-control signal is observed, the average run length can be computed (Xie *et al.*, 2001a):

$$ARL = \sum_{j=1}^{\infty} j\left(\prod_{i=1}^{I}(1-\alpha_i)\right)^{m} P_{j-mI} \tag{10.8}$$

where $m = \text{int}[j/I]$, the largest integer less than or equal to j/I.

In Table 10.1, some numerical values of average run lengths for different values of the adjustment interval I and trend slope β_1 are shown. The control limit coefficient k is set at 3 and average process starting level β_0 is set at 0. As expected, an increase in the slope will reduce the *ARL* significantly. On the other hand, although an increase in the adjustment interval will also reduce the *ARL*, it has a smaller effect on the *ARL*.

Table 10.1. Numerical values of *ARL* under different adjustment interval I and trend slope β_1

Adjustment interval I	Trend Slope β_1	*ARL*
10	0	369.41
10	0.01	363.13
10	0.05	251.84
10	0.1	111.64
10	0.2	21.71
1	0.01	369.30
5	0.01	367.73
10	0.01	363.13
20	0.01	346.01
50	0.01	258.63

10.2.2. An example with linear trend

The results in the previous section are general, a specific model for the trend is needed in further analysis. The linear trend, a common first order approximation will be investigated here. For tool-wear processes, for example, the slow wear at the beginning can usually be modeled by a linear model before a scheduled replacement takes place. The approach can be easily generated to other models.

Example 10.1:

A set of tool-wear data is used as an example. Table 10.2 contains 40 dimensional measurement errors produced in a face milling process (Bäckström and Wiklund, 1998). The data are collected consecutively in fixed time intervals between the changing of milling tools. A run plot of these data indicates the existence of a linear trend. Hence a linear regression model is used:

$$y_i = a + b \cdot i + \varepsilon_i \tag{10.9}$$

where y_i is the sample measurement error, a is the average starting level of the process, b is the slope of the linear trend and ε_i is the noise or random fluctuation, which is assumed normally distributed with zero mean and constant standard deviation σ .

In this case, $a = 1.485804$, $b = 0.113144$ and $\sigma = 0.506887$. By introducing $d = b/\sigma$, and using k-sigma limits, we have

$$\begin{aligned} \alpha_j &= P\{y_j > UCL\} + P\{y_j < LCL\} \\ &= 1 - \Phi\left(\frac{UCL - \mu_{y_j}}{\sigma_{y_j}}\right) + \Phi\left(\frac{LCL - \mu_{y_j}}{\sigma_{y_j}}\right) \\ &= \Phi(-k + dj) + \Phi(-k - dj) \end{aligned} \tag{10.10}$$

and

$$P_j = \alpha_j \cdot \frac{(1-\alpha_{j-1})}{\alpha_{j-1}} P_{j-1} \tag{10.11}$$

and the average run length is obtained as before.

Table 10.2. Dimensional measurement errors of the cutting tool in a face milling process (Bäckström and Wiklund, 1998)

Index	Error	Index	Error	Index	Error	Index	Error
1	1	11	2.98	21	4.04	31	4.46
2	1.04	12	3.08	22	4.06	32	5.92
3	1.6	13	3.22	23	4.08	33	4.92
4	2.26	14	2.88	24	4.46	34	6.3
5	1.7	15	2.86	25	4.08	35	5.06
6	2.54	16	3.12	26	5.06	36	5.7
7	2.42	17	3.34	27	4.34	37	5.42
8	2.76	18	4.2	28	4.44	38	5.7
9	2.9	19	3.34	29	2.9	39	7.74
10	2.88	20	3.74	30	5.5	40	6.98

Table 10.3 shows some numerical values of the average run lengths for different values of the adjustment interval I for this face milling process, each with control limits UCL and LCL set at 3 and –0.028 respectively. In generating the ARL table, approximation is used to stop the summation once the number of summation exceeds 400, and the differentiation between $ARL|_{j=1:J}$ truncated at value of J and $ARL|_{j=1:J-1}$ truncated at a value J-1 becomes smaller than a predefined infinitesimal τ, $\tau = 0.00001$.

□

Table 10.3. Some numerical values of *ARL* for the example

Adjustment interval I	ARL
1	314.78
5	82.51
10	15.75
20	11.06

10.2.3. The case of one-sided limit

Usually for trended process, the trend is known to be increasing or decreasing and the quality characteristic is also of smaller-the-better or larger-the-better type. A one sided limit can then be used. That is,

$$\alpha_j = P\{y_j > UCL\} \quad \text{or} \quad \alpha_j = P\{y_j < LCL\} \tag{10.12}$$

and here we assume that only the UCL is of interest. In this case, using again the linear trend model, we simply have

$$\alpha_j = P\{y_j > UCL\} = \Phi(-k + dj) \tag{10.13}$$

Furthermore, we have

$$P_j = \alpha_j \cdot \frac{(1-\alpha_{j-1})}{\alpha_{j-1}} P_{j-1} \tag{10.14}$$

as before. Based on these formulas, the average run length and other related quantities could be computed easily.

10.3. Modification of process capability indices

Process capability analysis is one of the most popular SPC tools (Kotz and Lovelace, 1998) and many indices are now an integral part of most commercial SPC software. It is commonly argued that for any process capability index to be meaningful, the process itself must be stable. For processes with unknown trend patterns, traditional process capability indices should not be used directly. On the other hand, it would be useful if an interpretation of process capability indices can be provided when the process is trended and regularly adjusted.

To study the capability of trended processes, Spiring (1991) considers the process capability to be dynamic rather than a constant. Such an interpretation, although reasonable, seems to be difficult to implement from a practical point of view. Furthermore, the paper proposes the loss-function focused index rather than the traditional variance-focused capability indices (Kane, 1986).

10.3.1. Interpretation of process capability indices

The traditional process capability index is defined as (Kane, 1986):

$$C_p = \frac{USL - LSL}{6\sigma} \tag{10.15}$$

where *USL, LSL,* and σ are the upper specification limit, the lower specification limit and the standard deviation of the characteristic, respectively. Another commonly used index is defined as (Kane, 1986)

$$C_{pk} = \frac{\min(USL - \mu, \mu - LSL)}{3\sigma} \tag{10.16}$$

where μ is the mean of the process and this capability index can deal with both process mean and variation.

Although the traditional capability indices should not be used directly when the process shows a certain trend, they can easily be modified by incorporating different sources of variation (Xie *et al.*, 2002).

Consider a series of measurement values, $(x_1, x_2, ..., x_n)$ where x_i is $N(\mu_i, \sigma)$. Since the process shows a certain trend, we assume that the expected value μ_i of the individual measurements depends on i. The standard deviation is assumed to remain constant and it is estimated or known. A process capability index can be defined as before with the mean defined as $\mu = \sum_{i=1}^{n} \mu_i / n$. On the other hand, because the mean is changing for each individual value, we should use the actual process variability as a measure of the process variation, denoted by σ_P^2 here. It is clear that

$$\sigma_P^2 = \sigma_T^2 + \sigma^2 \tag{10.17}$$

where σ_T^2 is the nonrandom variation caused by the trend component and σ^2 represents the variation of the individual measurement.

As σ_T^2 represents the variations of the trend component, a specific model is needed to measure its effect. We assume that there is a linear trend within the period before adjustment, that is,

$$x_i = a + bi + \varepsilon_i,\ i=1,2,\ \ldots,\ I \tag{10.18}$$

where a and b are the linear model parameters, ε is the error term with zero mean and a standard deviation of σ^2. Furthermore, I is the time a planned adjustment is to be made.

We assume that the process is adjusted periodically at every I time units before the replacement of tools or power supply according to a fixed schedule. By the usual partitioning of sum of squares, we can show that

$$\sigma_T^2 = \frac{\sum_{i=1}^{I} (a + bi - \mu)^2}{I} = \frac{\sum_{i=1}^{I} (a + bi)^2 - I \cdot \mu^2}{I} \tag{10.19}$$

Furthermore, when I is an even number, it is clear that $a + b \cdot I / 2 = \mu$. That is, the mid-value, or the value at $i = I / 2$, should be the same as the mean. In that case,

$$\sigma_T^2 = 2\frac{\sum_{i=I/2+1}^{I}(a+bi-\mu)^2}{I} \tag{10.20}$$
$$= \frac{2b^2}{I}\cdot\left[\frac{1}{6}\frac{I}{2}(\frac{I}{2}+1)(I+1)\right] = \frac{b^2}{6}(\frac{I}{2}+1)(I+1)$$

Note that when the adjustment interval I is large, which is usually the case when there is a slow trend,

$$\sigma_T^2 \approx \frac{b^2 I^2}{12} \quad \text{or} \quad \sigma_P^2 \approx \frac{b^2 I^2}{12} + \sigma^2 \tag{10.21}$$

Here, we can redefine the process capability indices after accounting for the total variability of the process, i.e.,

$$C_p^* = \frac{USL-LSL}{6\sigma_P} = \frac{USL-LSL}{6\sqrt{\sigma_T^2+\sigma^2}} \approx \frac{USL-LSL}{6\sqrt{b^2I^2/12+\sigma^2}} \tag{10.22}$$

and

$$C_{pk}^* = \frac{\min(USL-\mu, \mu-LSL)}{3\sigma_P} \approx \frac{\min(USL-\mu\ , \mu-LSL)}{3\sqrt{b^2I^2/12+\sigma^2}} \tag{10.23}$$

For comparison, Table 10.4 displays the values of C_p^* under typical situations. Without loss of generality, $USL-LSL$ is set to be equal to 6, with σ and $b\cdot I$ varying within certain range. Note that when b or I is much smaller than σ^2, $C_p^* \approx C_p$, which is the same as the traditional C_p for a stable process.

It can be seen that the factor $1/\sqrt{b^2I^2/(12\sigma^2)+1}$ is less than or equal to 1. That is, the traditional C_p is the maximum achievable process capability through trend adjustment. In general, for any fix value of σ, $C_p^* \le C_p$.

Table 10.4. Some numerical values of process capability index C_p^*

	bI						
σ	0	0.2	0.4	0.8	1.6	3.2	6.4
0.3	3.3333	3.2733	3.1109	2.6414	1.8157	1.0296	0.5343
0.4	2.5	2.4744	2.4019	2.1651	1.6366	0.9934	0.5290
0.5	2	1.9868	1.9487	1.8157	1.4691	0.9520	0.5225
0.6	1.6667	1.6590	1.6366	1.5555	1.3207	0.9078	0.5148
0.7	1.4286	1.4237	1.4095	1.3566	1.1924	0.8628	0.5062
0.8	1.25	1.2468	1.2372	1.201	1.0825	0.8184	0.4967
0.9	1.1111	1.1088	1.1021	1.0762	0.9885	0.7754	0.4866
1	1	0.9983	0.9934	0.9744	0.9078	0.7346	0.4760
1.1	0.9091	0.9078	0.9041	0.8897	0.8382	0.6962	0.4651
1.2	0.8333	0.8324	0.8295	0.8183	0.7777	0.6603	0.4539
1.3	0.7692	0.7685	0.7662	0.7574	0.7248	0.6270	0.4427
1.4	0.7143	0.7137	0.7119	0.7048	0.6783	0.5962	0.4314

10.3.2. Adjustment interval meeting capability requirement

An important application of this study is the determination of adjustment time. Suppose a certain minimum level of capability is given, say, c_p or c_{pk}. The maximum process adjustment time can be determined as follows

$$\frac{USL-LSL}{6\sqrt{b^2I^2/12+\sigma^2}} \geq c_p \tag{10.24}$$

and the adjustment interval meeting this requirement is given by

$$I \leq 2\sqrt{3}\cdot\left[\left(\frac{USL-LSL}{6c_p}\right)^2-\sigma^2\right]^{1/2}/b \tag{10.25}$$

With a C_{pk} requirement, we have in a similar way that

$$\frac{\min(USL-\mu,\mu-LSL)}{3\sqrt{b^2I^2/12+\sigma^2}}\geq c_{pk} \tag{10.26}$$

which gives

$$I\leq 2\sqrt{3}\left[\left(\frac{\min(USL-\mu,\mu-LSL)}{3c_{pk}}\right)^2-\sigma^2\right]^{1/2}/b \tag{10.27}$$

Note that to be able to meet the process capability requirement by adjusting for the trend, the potential capability should be greater than the requirement and C_p is the maximum achievable capability.

Example 10.2:

Suppose that for a certain process, the trend has been estimated as

$$\hat{x}_i=0.003135+0.000552\cdot i+\varepsilon_i$$

where $\varepsilon_i\sim N(0,0.0034)$. That is, $\hat{b}$=0.000552 and $\hat{\sigma}=0.0034$. Furthermore, suppose that *USL*=0.04, *LSL*=0 and $\mu=0.02$. Also, assume that the initial adjustment interval is *I*=40. The value of $\hat{C}_p^*=\hat{C}_{pk}^*$ is 0.923. This can be considered to be too low, although the maximum achievable capability $\hat{C}_p$=1.96 is acceptable.

Suppose that there is a process capability requirement of c_p^*=1.5 and this has to be met with proper process adjustment. In this case, since $C_p=1.96$, which is greater than the required capability value, the requirement can be met by only changing the adjustment interval. The appropriate adjustment interval can be computed to be *I*=27.89. If a value is taken every hour, then an adjustment should be scheduled every 27.89 hours.

□

10.4. Cost consideration and decision making

Usually for trended processes, adjustments are necessary. Adjustment should be carried out when the process is out-of-control and also when some other conditions for prescribed maintenance are met. For this type of processes, it is important to be able to determine not only the optimal control limits, but also the adjustment time from an economic point of view. We present here some basic ideas and preliminary results.

10.4.1. Process failure and adjustment model

A justification of using Poisson failure model is explained first. In applying economic design, knowledge on the distribution of the process failure mechanism is required. The Poisson model is often assumed for such a purpose, where the process is assumed to fail after the first outside shock and hence is referred to as subject to single process disruption in Lorenzen and Vance (1986). The out-of-control shocks are modeled as a Poisson process and the distribution of the number of shocks is:

$$f_x(x:\lambda)=\frac{e^{\lambda}\cdot\lambda^{x}}{x!},\quad x=0,1,2,\ldots \tag{10.28}$$

for which the mean is λ. This implies that the failure time distribution of an exponential, or a constant hazard rate.

However, for processes subjected to increasing hazard rate, the Weibull model is preferred. As seen in Banerjee and Rahim (1988), an economic design of X bar control chart under the Weibull failure mechanism is studied for systems with increasing hazard rate. When an economic design model is to be used for this type of processes, because the process is subject to routinely made adjustments, the Poisson failure model can still be justified.

For example, with routine preventive adjustments, the process is saw-tooth shaped and consists of many production cycles. For a process of this type, the wear-out factor is compensated by preventive adjustment. Hence, the hazard function concerned here is not closely related to the wearing out process because of frequent adjustments. In this case, the actual failure can still be adequately represented using a Poisson model, which will be used in the following.

As the name implied, an engineering controlled drifting process is subject to routinely made adjustments. The routine adjustments are carried out in a way as if there is a dead band placed around the process mean. As shown in Figure 10.3, the monitoring chart for such a process is divided into three areas:

(a) the central area inside the dead band,

(b) the gray area outside the dead band but inside the control limits (*UCL* and *LCL*), and

(3) the failure area outside the control limits.

When sample points are found in the central area, the process is left free of adjustment. When a sample point is found in the gray area, the process is regularly adjusted with no searching of assignable causes. Such adjustment is referred to as preventive adjustment. Finally, when a sample point is found in the failure area, the process is stopped for corrective adjustment with assignable cause found and eliminated accordingly. This control scheme is more related to statistical process control procedures with which the process is not adjusted unless there is a strong statistical evidence that the process mean or variability has changed.

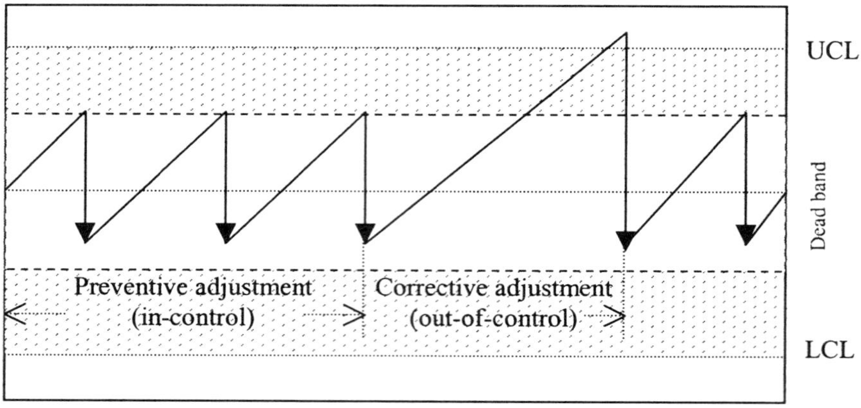

Figure 10.3. A trended and regularly adjusted process with its monitoring and adjustment policy.

10.4.2. Cost model

Both adjustment frequency and control chart design which include the control limits, sampling frequency and sample size, can be considered from economic point of view. In terms of economic design, the determination of any of these factors relies on the finding of minimum value of the expected total cost.

As the underlying failure mechanism for engineering controlled drifting process is of the Poisson model, the economic design can be made based on the unified cost model of Lorenzen and Vance (1986), where the process in question is assumed to be in a renewal cycle consisting of the following four states:

(1) In-control State, where the process output follows a distribution with mean $\overline{x}_i = x_a$ and standard deviation σ_x. ($L<x_a<U$ is the acceptable range);

(2) Out-of-control State, where the mean goes beyond the acceptable range of the control chart;

(3) Searching State, when searching for assignable causes; and

(4) Adjustment State, when the process is being adjusted.

There are six types of cost under consideration: (i) sampling cost, (ii) cost of the process when it is running in an out-of-control state, (iii) cost of process when it is running in an in-control state, (iv) cost of false alarms, (v) cost of searching for assignable cause when process is out-of-control and (vi) adjustment cost. A list of the notations used in the definition of these cost components is as follows:

λ: 1/mean time process is in control

v: Production rate

C_0: Average cost per unit when the production process is in control

C_1: Average cost per unit when the production process is out of control

C_a: Average cost per intermediate adjustment

C_f: Average cost per false alarm

C_r: Average cost to locate the assignable causes

C_s: Sampling and inspection cost per unit

a: Inspection and sampling cost per sample

Y: Cost per time of false alarm

W: Cost per time of process reset

The total expected cost is the sum of these six components:

$$E(C) = E(C_0) + E(C_1) + E(C_s) + E(C_f) + E(C_r) + E(C_a)$$

With a specific process model, the expected total cost per unit time can be obtained. A number of assumptions are made:

(a) The control chart is used to maintain current surveillance of a process that is subjected to a single assignable cause, and occurrence of assignable cause is assumed exponential distributed with parameter λ (as in Poisson model);

(b) The time of taking a sample, inspecting and charting, and making regular adjustment are negligible since normally they should and have to be done by machine automatically;

(c) The sample size equals to one, which is usually the case for automated production process;

(d) The preventive adjustment is administered at every equally-spaced time interval I.

The expected total cost per unit $E(C)$ is given by (Cai *et al.* 2002):

$$E(C) = \frac{C_0 \times E(n_1) + C_1 \times E(n_2) + Y \times E(n_f) + W}{E(N)} + a + \frac{C_a}{I} \tag{10.29}$$

where:

$$E(C_0) = C_0 \times E(n_1) / E(N)$$
$$E(C_1) = C_1 \times E(n_2) / E(N)$$
$$E(C_s) = a$$
$$E(C_f) = Y \times E(n_f) / E(N)$$
$$E(C_r) = W / E(N)$$
$$E(C_1) = C_a / I$$

Let n_1 and n_2 be the average number of units produced during 'in-control' state and 'out-of–control' state for each renewal cycle, respectively. Let n_f be the average number of false alarms per cycle.

Since the 'in-control' time is exponentially distributed with parameter λ, we have

$$\begin{aligned}\text{Prob}(n_1 = i) &= \text{Prob}(i/v \le x < (i+1)/v) \\ &= \int_{i/v}^{(1+i)/v} \lambda e^{-\lambda x} dx = e^{-\lambda i/v} - e^{-\lambda(i+1)/v}\end{aligned} \tag{10.30}$$

so that

$$E(n_1) = \sum_{i=0}^{+\infty} i \times \left(e^{-\lambda i/v} - e^{-\lambda(i+1)/v}\right) = 1/(e^{\lambda/v} - 1) \tag{10.31}$$

and it can also be obtained that (Cai *et al.* 2000),

$$\begin{aligned}E(N) &= E(n_1 + n_2) = E(n_1) + E(n_2) \\ &= \frac{1}{e^{\lambda/v} - 1} + \frac{1}{1-\beta} = \frac{e^{\lambda/v} - \beta}{(e^{\lambda/v} - 1)(1-\beta)}\end{aligned} \tag{10.32}$$

The overall process expected unit cost can then be obtained as

$$\begin{aligned}E(C) &= \frac{C_0/(e^{\lambda/v} - 1) + C_1/(1-\beta) + Y\alpha\big/(e^{\lambda/v} - 1) + W}{\dfrac{e^{\lambda/v} - \beta}{(e^{\lambda/v} - 1)(1-\beta)}} + a + \frac{C_a}{I} \\ &= \frac{[(e^{\lambda/v} - 1)W + Y\alpha + (C_0 - C_1)] \times (1-\beta)}{e^{\lambda/v} - \beta} + a + C_1 + \frac{C_a}{I}\end{aligned}$$

The optimal adjustment interval can then be obtained by minimizing $E(C)$.

Example 10.3:

As an example *E(C)* curves of three monotonically drifting processes with $x_{i1} = b \cdot i$, $x_{i2} = e^{b \cdot i}$, and $x_{i3} = b \cdot i + \nabla|_{i>t}$, are shown in the Figure 10.4. The three curves are all convex with a local minimum pointing to the lowest overall cost.

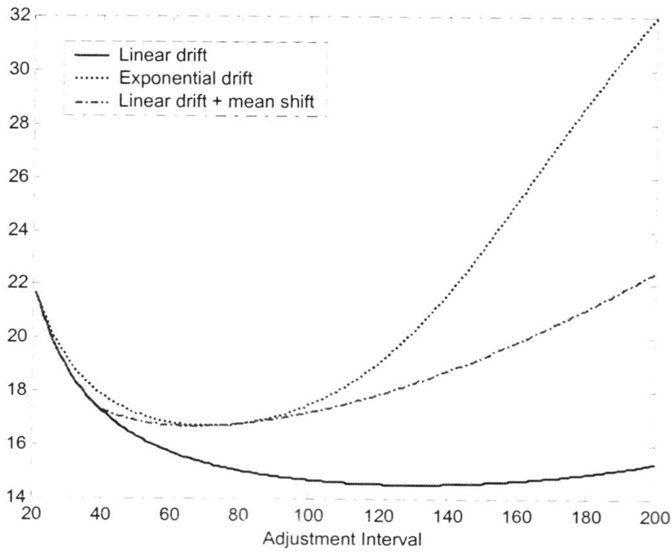

Figure 10.4. Expected cost curve for some trended process.

The cost parameters in this cost model are assigned as in the following, $C_0 = \$10$, $C_1 = \$50$, $Y = \$200$, $W = \$100$, $a = \$1$, $C_a = \$200$, $\lambda = 0.01$, and $v = 1$. For any assumed value of *I*, the size of acceptable range would become $d = (b_I \times I)/\sigma$. The minimum of the expected unit cost *E(C)* is reached when $I = 29$ and $k = 13$ and the minimum cost is 19.0687. This suggests a tool change at every $I = 29$ for this particular milling operation under Shewhart chart monitoring of $k = 13$.

□

10.4.3. Further discussions

As cost is an important issue, especially in the case of trended and adjusted processes, more studies should focus on combined statistical properties and cost considerations. Usually where SPC is concerned, only the statistical issues and properties are discussed. Economic design in the past also mainly involves SPC related decision variables, not process adjustment issues.

On the other hand, since regular adjustment is important for trended processes, the adjustment frequency is as important from the managerial point of view as they will affect costs. It is not easy to consider all aspects of the process adjustment and monitoring. This poses many interesting problems for further research. Different cost models and adjustment strategies can be used and the sensitivity of the model parameters should be investigated.

There are numerous books and articles on engineering process control, and some that discuss the integration of engineering process control and statistical process control are Box and Luceno (1997), Montgomery *et al.* (1994) and Gob *et al.* (1999). It would be of interest to study the process capability issues under different adjustment strategies since process capability indices, if used properly, could provide some interesting summary information about the process when customer specifications have to be considered.

Another practical issue involves the estimation of the parameters of process model. A linear model for the trend is usually a first order approximation but quadratic and other forms could be possible. When limited data is available, the estimation error will also have a significant impact on the chart design and overall cost.

References

Acosta-Mejia, C.A. (1999). Improved *p* charts to monitor process quality. *IIE Transactions*, **31**, 509-516.

Alwan, L.C. (1992). Effects of autocorrelation on control charts. *Communication in Statistics – A: Theory and Methods*, **21**, 1025-1049.

Alwan, L.C. and Roberts, H.V. (1988). Time series modelling for statistical process control. *Journal of Business and Economic Statistics*, **6**, 67-75.

Amin, R.W. and Widmaier, O. (1999). Sign control charts with variable sampling intervals. *Communication in Statistics – A: Theory and Methods*, **28**, 1961-1985.

Anscombe, F.J. (1948). The transformation of Poisson, binomial, and negative binomial data. *Biometrika*, **35**, 246-254.

Aparisi, F. and Haro, C.L. (2001). Hotelling's T-2 control chart with variable sampling intervals. *International Journal of Production Research*, **39**, 3127-3140.

Atienza, O.O.; Tang, L.C. and Ang, B.W. (2000). A uniformly most powerful cumulative sum scheme based on symmetry. *The Statistician*, **49**, 209-217.

Backstrom, M. and Wiklund, H. (1998). Development of a multi-tooth approach to tool condition monitoring in milling. *Insight*, **40**, 548-552.

Bai, D.S. and Lee, K.T. (1998). An economic design of variable sampling interval X control charts. *International Journal of Production Economics*, **54**, 57-64.

Banerjee, P.K. and M. A. Rahim, (1988). Economic design of X-bar control charts under Weibull shock models, *Technometrics*, **30**, 407-414.

Banks, J. (1989). *Principles of Quality Control*. Wiley, New York.

Besterfield, D.H. (1997). *Quality Control*. Prentice-Hall, New Jersey.

Bhat, U.N. and Lal, R. (1988). Number of successes in Markov trials, *Advances in Applied Probability*, **20**, 677-680.

Bhat, U.N. and Lal, R. (1990). Attribute control charts for Markov dependent production processes. *IIE Transactions*, **22**, 181-188.

Bohning, D. (1998). Zero-inflated Poisson models and C.A.MAN: A tutorial collection of evidence. *Biometrical Journal*, **40**, 833-843.

Borror, C.M.; Champ, C.W. and Rigdon, S.E. (1998). Poisson EWMA control charts, *Journal of Quality Technology*, **30**, 352-361.

Bourke, P.D. (1991). Detecting shift in fraction nonconforming using run-length control charts with 100% inspection. *Journal of Quality Technology*, **23**, 225-238.

Bourke, P.D. (2001). Sample size and the Binomial CUSUM control chart: the case of 100% inspection. *Metrika*, **53**, 51-70.

Box, G.E.P. and Cox, D. R. (1964). An analysis of transformations. *Journal of the Royal Statistical Society - Series B*, **26**, 211-252.

Box, G.E.P.; Jenkins, G.M. and Reinsel, G.C. (1994). *Time Series Analysis, Forecasting and Control*. Prentice-Hall, New Jersey.

Box, G.E.P. and Luceno, A. (1997). *Statistical Control by Monitoring and Feedback Adjustment*. Wiley, New York.

Braun, W.J. (1999). Run length distributions for estimated attributes charts. *Metrika*, **50**, 121-129.

Broadbent, S.R. (1958). The inspection of a Markov process. *Journal of Royal Statistical Society*, **20**, 111-119.

Brook, D. and Evans, D.A. (1972). An approach to the probability distribution of CUSUM run length. *Biometrika*, **59**, 539-549.

Burke, R.J.; Davis, R.D.; Kaminsky, F.C. and Roberts, A.E.P. (1995). The effects of inspector errors on the true fraction nonconforming: An industrial experiment. *Quality Engineering*, **7**, 543-550.

Cai, D.Q.; Xie, M. and Goh, T.N. (2001). SPC in an automated manufacturing environment. *International Journal of Computer-Integrated Manufacturing*, **14**, 206-211.

Cai, D.Q., Xie, M., Goh, T.N. and Tang, X.Y. (2002). Economic design for trended processes. *International Journal of Production Economics*, to appear.

Calvin, T.W. (1983). Quality control techniques for 'zero-defects'. *IEEE Trans. on Components, Hybrids, and Manufacturing Technology*, **6**, 323-328.

Case, K. (1980). The *p* control chart under inspection error. *Journal of Quality Technology*, **12**, 1-9.

Champ, C.W. and Woodall, W.H. (1987). Exact results for Shewhart control charts with supplementary run rules. *Technometrics*, **29**, 393-399.

Chan, L.Y.; Xie, M. and Goh, T.N. (1997). Two-stage control charts for high yield processes. *International Journal of Reliability, Quality and Safety Engineering*, **4**, 149-165.

Chan, L.Y.; Xie, M. and Goh, T.N. (2000). A cumulative quantity control chart for manufacturing processes. *International Journal of Production Research*, **38**, 397-408.

Chang, T.C. and Gan, F.F. (1999). Charting techniques for monitoring a random shock process. *Quality and Reliability Engineering International*, **15**, 295-301.

Chen, G. (1997). The mean and standard deviation of the run length distribution of $\overline{X}$ charts when control limits are estimated. *Statistica Sinica,* **7**, 789-798.

Chen, G. (1998). The run length distributions of R, s and s^2 control charts when σ is estimated. *Canadian Journal of Statistics,* **26**, 311-322.

Chen, G.M.; Cheng, S.W. and Xie, H.S. (2001). Monitoring process mean and variability with one EWMA chart. *Journal of Quality Technology*, **33**, 223-233.

Chiu, W.K. (1975). Economic design of attribute control charts. *Technometrics*, **17**, 81-87.

Chou, C.Y.; Liu, H.R. and Chen, C.H. (2001). Economic design of averages control charts for monitoring a process with correlated samples, *International Journal of Advanced Manufacturing Technology* **18**, 49-53.

Chou, Y.M., Mason, R.L. and Young, J.C. (1999). Power comparisons for a Hotelling's T-2 statistic. *Communication in Statistics – A: Theory and Methods*, **28**, 1031-1050.

Chung, K.J. (1991). A simplified procedure for the economic design of control charts: A unified approach. *Engineering Optimization*, **17**, 313-320.

Chung, K.J. (1992). An efficient procedure for the economic design of *np* control charts. *International Journal of Quality and Reliability Management*, **9**, 58-68.

Chung, K.J. (1995). An algorithm for the economic design of *np* charts for a multiplicity of assignable causes. *Journal of the Operational Research Society*, **46**, 1374-1385.

Consul, P.C. (1989). *Generalized Poisson Distributions: Properties and Applications*. Marcel Dekker, New York.

Costa, A.F.B. (1993). Joint economic design of x-bar and R control charts for processes subject to two independent assignable causes. *IIE Transactions*, **25**, 27-33.

Costa, A.F.B. and Rahim, M.A. (2000). Economic design of (X)over-bar and R charts under Weibull shock models. *Quality and Reliability Engineering International*, **16**, 143-156.

Crowder, S.V. (1987). A simple method for studying run-length distribution of exponentially weighted moving average charts. *Technometrics*, **29**, 123-129.

Derman, C. and Ross, S.M. (1997). *Statistical Aspects of Quality Control.* Academic Press, San Diego.

Doganaksoy, N.; Faltin, F.W. and Tucker, W.T. (1991). Identification of out-of-control quality characteristics in a multivariate manufacturing environment. *Communications in Statistics – A: Theory and Methods*, **20**, 2775-2790.

Duncan, A.J. (1986). The economic design of x-bar charts used to maintain current control of a process. *Journal of the American Statistical Association*, **51**, 228-242.

English, J.R. (1994). Fixed control limits for continuous flow processes. *IEEE Transactions on Engineering Management*, **41**, 310-314.

Ermer, D.S. (1995). Advanced SPC for high quality electronic card manufacturing. *Quality Engineering*, **8**, 283-299.

Freund, R. A. (1957). Acceptance control charts. *Industrial Quality Control*, **14**, 13–23.

Gan, F.F. (1990). Monitoring Poisson observations using modified exponentially weighted moving average control charts. *Communications in Statistics – B: Simulation and Computation*, **19**, 103-124.

Gan, F.F. (1991). An optimal design of CUSUM quality control charts. *Journal of Quality Technology*, **23**, 279-286.

Gan, F.F. (1993). An optimal design of CUSUM control charts for binomial counts. *Journal of Applied Statistics*, **20**, 445-458.

Gardiner, J.S. (1987). A Note on the average run length of cumulative sum control charts for count data. *Quality and Reliability Engineering International*, **3**, 53-55.

Glushkovsky, E.A. (1994). On-line G-control chart for attribute data. *Quality and Reliability Engineering International*, **10,** 217-227.

Gob, R.; Del Castillo, E. and Drager, K. (1999). Run length analysis of Shewhart charts applied to drifting processes under an integrative SPC/EPC model. *Metrika*, **50**, 137-16.

Goel, A.L. and Wu, S.M. (1971). Determination of ARL and a contour nomogram for CUSUM charts to control normal mean. *Technometrics*, **13**, 221-230.

Goh, T.N. (1987). A control chart for very high yield processes, *Quality Assurance*, 13, 18-22.

Goh, T.N. (1993). Some practical issues in the assessment of nonconforming rates in a manufacturing process. *International Journal of Production Economics*, **33**, 81-88.

Goh, T.N. and Xie, M. (1994). A new approach to quality in a near zero-defect Environment. *Total Quality Management*, **12**, 241-250.

Goh, T.N. and Xie, M. (1995). Statistical process control for low nonconformity processes. *International Journal of Reliability, Quality and Safety Engineering*, **2**, 15-22.

Grant, E.L. and Leavenworth, R.S. (1996). *Statistical Quality Control.* McGraw-Hill, New York.

Hawkins, D.M. (1981). A CUSUM for a scale parameter. *Journal of Quality Technology*, **13**, 102-110.

Hayter, A.J. and Tsui, K.L. (1994). Identification and quantification in multivariate quality control problems. *Journal of Quality Technology*, **26**, 197-207.

Ho, C. and Case, K.E. (1994). Economic design of control charts: a literature review. *Journal of Quality Technology*, **26**, 39-53.

Jiang, W.; Tsui, K.L. and Woodall, W.H. (2000). A new SPC monitoring method: The ARMA chart. *Technometrics*, **42**, 399-410.

Johnson, N.L. and Kotz, S. (1969). *Discrete Distributions*. Wiley, New York.

Johnson, N.L. and Kotz, S. and Balakrishnan, N. (1970). *Continuous Univariate Distributions*. Wiley, New York.

Johnson, N.L.; Kotz, S. and Wu, X. (1991). *Inspection Errors for Attributes in Quality Control*. Chapman & Hall, London.

Jolayemi, J.K. (2000). An optimal design of multiattribute control charts for processes subject to a multiplicity of assignable causes. *Applied Mathematics and Computation*, **114**, 187-203.

Jones, L.A., Woodall, W.H. and Conerly, M.D. (1999). Exact properties of demerit control charts. *Journal of Quality Technology*, **31**, 207-216.

Kaminsky, F.C.; Benneyan, R.D.; Davis, R.D. and Burke, R.J. (1992). Statistical control charts based on a geometric distribution. *Journal of Quality Technology*, **24**, 63-69.

Kane, V. (1986). Process capability indices. *Journal of Quality Technology*, **18**, 41-52.

Kanji, G.K. and Arif, O.H. (2001). Median rankit control chart for Weibull distribution. *Total Quality Management*, **12**, 629-642.

Kotz, S. and Lovelace, C.R. (1998). *Process Capability Indices in Theory and Practice*. Oxford University Press, London.

Kramer, H. and Schmid, W. (1997). Control charts for time series. *Nonlinear Analysis: Theory, Methods & Applications*, **30**, 4007-4016.

Kuralmani, V.; Xie, M.; Goh, T.N. and Gan, F.F. (2002). A conditional decision procedure for high yield processes, *IIE Transactions*, to appear.

Lai, C.D.; Xie, M. and Govindaraju, K. (2000). Study of a Markov model for a high-quality dependent process. *Journal of Applied Statistics*, **27**, 461-473.

Liu, R.Y. (1995). Control charts for the multivariate processes. *Journal of the American Statistical Association*, **90**, 1380-1387.

Lo, V.S.Y. (1994). On estimation methods of input parameters for economic attribute control charts. *Statistician*, **43**, 439-451.

Lorenzen, T.J. and Vance, L.C. (1986). The economic design of control charts: A unified approach. *Technometrics*, **28**, 3-10.

Lowery, C.A., and Montgomery, D.C. (1995). A review of multivariate control charts. *IIE Transactions,* **27**, 800-810.

Lu, X.S. (1999). *Statistical Control Charts for Automated and High-quality Manufacturing Processes*. PhD Dissertation, National University of Singapore.

Lu, X.S.; Xie, M. and Goh, T.N. (1998). Control chart for multivariate attribute processes. *International Journal of Production Research*, **36**, 3477-3489.

Lu, X.S.; Xie, M. and Goh, T.N. (2000). An investigation of the effects of inspection error on the run-length control charts. *Communications in Statistics – B: Simulation and Computation*, **29**, 315-335.

Lucas, J.M. (1985). Counted data CUSUM's. *Technometrics*, **27**, 129-144.

Lucas, J.M. (1989). Control schemes for low count levels. *Journal of Quality Technology*, **21**, 199-201.

Lucas, J.M. and Saccucci, M.S. (1990). Exponentially weighted moving average control schemes. *Technometrics*, **24**, 34-45.

Luceno, A. and Puig-Pey, J. (2000). Evaluation of the run-length probability distribution for CUSUM charts: Assessing chart performance. *Technometrics*, **42**, 411-416.

MacGregor, J.F. and Harris, T.J. (1993). The exponentially weighted moving average control charts. *Journal of Quality Technology*, **25**, 106-118.

MacGregor, J.F. and Kourti, T. (1995). Statistical process control of multivariate processes. *Control Engineering Practice*, **3**, 403-414.

Madsen, R.W. (1993). Generalized binomial distributions. *Communications in Statistics – A: Theory and Methods*, **22**, 3065-3086.

Marcucci, M. (1985). Monitoring multinomial processes. *Journal of Quality Technology*, **17**, 86-91.

Mason, R.L.; Tracy, N.D. and Young, J.C. (1995). Decomposition of T^2 for multivariate control chart interpretation. *Journal of Quality Technology*, **27**, 99-108.

Mason, R.L. and Young, J.C. (1998). Why multivariate statistical process control? *Quality Progress*, **31**, 88-93.

McCool, J.I. and Joyner-Motley. T. (1998). Control charts applicable when the fraction nonconforming is small. *Journal of Quality Technology*, **30**, 240-247.

McShane, L.M. and Turnbull, B.W. (1991). Probability limits on outgoing quality for continuous sampling plans. *Technometrics*, **33**, 393-404.

McWilliams, T.P. (1994). Economic, statistical, and economic-statistical $\overline{X}$ chart designs. *Journal of Quality Technology*, **26**, 227-238.

McWilliams, T.P. (1996). Economic control charts: Relating the model and finding the optimal design. In *Statistical Applications in Process Control*, Eds. J.B. Keats and D.C. and Montgomery, Marcel Dekker, New York.

Messina, W.S. (1988). *Statistical Quality Control for Manufacturing Managers*. Wiley, New York.

Mitra, A. (1998). *Fundamentals of Quality Control and Improvement*. Prentice Hall, New Jersey.

Montgomery, D.C. (1980). The economic design of control charts: A review and literature survey. *Journal of Quality Technology*, **12**, 75-87.

Montgomery, D.C. (1982). Economic design of an x-bar control charts. *Journal of Quality Technology*, **14**, 40-43.

Montgomery, D.C. (2001). *Introduction to Statistical Quality Control.* Wiley, New York.

Montgomery, D.C. and Mastrangelo, C.M. (1991). Some statistical process control methods for autocorrelated data. *Journal of Quality Technology*, **23**, 179-193.

Montgomery, D.C.; Keats, J.B.; Runger, G.C. and Messina, W. S. (1994). Integrating statistical process control and engineering process control., *Journal of Quality Technology,* **26**, 79-87.

Moustakides, G.V. (1986). Optimal stopping times for detecting changes in distributions. *Annals of Statistics*, **14**, 1379-1387.

Nedumaran, G. and Pignatiello, J.J. (1999). On constructing T-2 control charts for on-line process monitoring. *IIE Transactions*, **31**, 529-536.

Nelson, L.S. (1984). The Shewhart control chart-tests for special causes. *Journal of Quality Technology*, **16**, 237-239.

Nelson, L.S. (1987). A chi-square control chart for several populations. *Journal of Quality Technology*, **19**, 229-240.

Nelson, L.S. (1994). Control chart for parts-per-million conforming items. *Journal of Quality Technology*, **26**, 239-240.

Oakland, J.S. (1986). *Statistical Process Control.* Heinemann Newnes, London.

Ohta, H.; Kusukawa, E. and Rahim, A. (2001). A *CCC-r* chart for high-yield processes. *Quality and Reliability Engineering International*, **17**, 439-446.

Page, E. S. (1954). Continuous inspection schemes. *Biometrika*, **24**, 199-205.

Page, E. S. (1961). Cumulative sum control charts. *Technometrics*, **3**, 1-9.

Patel, H.I. (1973). Quality control methods for multivariate binomial and Poisson distributions. *Technometrics*, **15**, 103-112.

Prabhu, S.S.; Runger, G.C. and Montgomery, D.C. (1997). Selection of the subgroup size and sampling interval for a CUSUM control chart. *IIE Transactions*, **29,** 451-457.

Quesenberry, C.P. (1995). Geometric *Q* charts for high quality processes. *Journal of Quality Technology*, **27,** 304-313.

Quesenberry, C.P. (1997). *SPC Methods for Quality Improvement.* Wiley, New York.

Radaelli, G. (1994). Poisson and negative binomial distribution dynamics for counted data under CUSUM-type charts. *Journal of Applied Statistics*, **21**, 347-356.

Reynolds, M.R. (1975). Approximations to the average run length in cumulative sum control charts. *Technometrics*, **17**, 65-71.

Reynolds, M.R. and Arnold, J.C. (2001). EWMA control charts with variable sample sizes and variable sampling intervals. *IIE Transactions*, **33**, 511-530.

Reynolds, M.R. and Stoumbos, Z.G. (2000). A general approach to modeling CUSUM charts for a proportion. *IIE Transactions*, **32**, 515-535.

Roberts, S.W. (1959). Control chart tests based on geometric moving averages. *Technometrics*, **1**, 239-250.

Robinson, P.B. and Ho. T.Y. (1978). Average run lengths of geometric moving average charts by numerical methods, *Technometrics*, **20**, 85-93.

Runger, G.C.; Alt, F.B. and Montgomery, D.C. (1996). Contributors to a multivariate statistical process control chart signal. *Communications in Statistics - A: Theory and Methods*, **25**, 2203-2213.

Ryan, T.P. (1989). *Statistical Methods for Quality Improvement.* Wiley, New York.

Ryan, T.P. and Schwertman, N. C. (1997). Optimal limits for attributes control charts. *Journal of Quality Technology*, **29**, 86-96.

Shewhart, W.A. (1926). Quality control charts. *Bell Systems and Technology Journal*, October, 593-603.

Shewhart, W.A. (1931). *Economic Control of Quality of Manufactured Product.* Van Nostrand, New York.

Shore, H. (2000). General control charts for attributes. *IIE Transactions*, **32**, 1149-1160.

Spiring, F.A. (1991). Assessing process capability in the presence of systematic assignable cause. *Journal of Quality Technology*, **23**, 125-134.

Spiring, F.A. and Cheng, S.W. (1998). An alternate variables control chart: The univariate and multivariate case. *Statistica Sinica*, **8**, 273-287.

Stimson, W.A. and Masterangelo, C.M. (1996). Monitoring serially-dependent process with attributes data, *Journal of Quality Technology*, **28**, 279-288.

Suich, R. (1988). The *c* control chart under inspection error. *Journal of Quality Technology*, **20**, 263-266.

Sun, J. and Zhang, G.X. (2000). Control charts based on the number of consecutive conforming items between two successive nonconforming items for the near zero-nonconformity processes. *Total Quality Management*, **11**, 235-250.

Surtihadi, J. and Raghavachari, M. (1994). Exact economic design of x-over-bar charts for general time in-control distributions. *International Journal of Production Research*, **32**, 2287-2302.

Tagaras, G. (1994). A dynamic programming approach to the economic design of x charts. *IIE Transactions*, **26**, 48-56.

Tagaras, G. and Lee, H.L. (1988). Economic design of control charts with different control limits for different assignable causes. *Management Science*, **34**, 1347-1366.

Tang, X.Y. (2000). *The Economic Design of Control Charts for High Quality Processes*. PhD Dissertation, National University of Singapore.

Tang, X.Y.; Xie, M. and Goh, T.N. (2000). A note on economic-statistical design of cumulative count of conforming control chart. *Economic Quality Control*, **15**, 3-14.

van Brackle, L.N. and Reynolds, M.R. (1997). EWMA and CUSUM control charts in the presence of correlation. *Communications in Statistics – B: Computation and Simulation*, **26**, 979-1008.

von Collani, E. (1988). Economically optimal *c* and *np* -control charts. *Metrika*, **36**, 215-232.

van Dobben de Bruyn, C.S. (1968). *Cumulative Sum Tests*. Griffin, London.

Vellaisamy, P. and Sankar, S. (2001). Sequential and systematic sampling plans for the Markov-dependent production process. *Naval Research Logistics*, **48**, 451-467.

Wade, M.R. and Woodall, W.H. (1993). A review and analysis of cause-selecting control charts. *Journal of Quality Technology*, **25**, 161-169.

Wasserman, G.S. (1995). An adaptation of the EWMA chart for short-run SPC. *International Journal of Production Research*, **33**, 2821-2833.

Wetherill, G.B. and Brown, D.W. (1991). *Statistical Process Control-Theory and Practice*. Chapman & Hall, London.

Wheeler, D.J. and Chambers, D.S. (1992). *Understanding Statistical Process Control*. SPC Press, Knoxville, TN.

White, C.H. and Keats, J.B. (1996). ARLs and higher-order run-length moments for the Poisson CUSUM. *Journal of Quality Technology*, **28**, 363-369.

Wierda, S.J. (1994). Multivariate statistical process control-recent results and directions for future research. *Statistica Neerlandica*, **48**, 147-168.

Winterbottom, A. (1993). Simple adjustments to improve control limits on attribute charts. *Quality and Reliability Engineering International*, **9**, 105-109.

Woodall, W.H. (1985). The statistical design of quality control charts. *The Statistician*, **34**, 155-160.

Woodall, W.H. (1997). Control charts based on attribute data: bibliography and review. *Journal of Quality Technology*, **29**, 172-184.

Woodall, W.H. and Adams, B.M. (1993). The statistical design of CUSUM charts. *Quality Engineering*, **5**, 559-570.

Wu, Z. and; Spedding, T.A. (1999). Evaluation of ATS for CRL control chart. *Process Control & Quality*, **11**, 183-191.

Wu, Z.; Yeo, S.H. and Fan, H.T. (2000). A comparative study of the CRL-type control charts. *Quality and Reliability Engineering International*, **16**, 269-279.

Wu, Z.; Zhang, X.L. and Yeo, S.H. (2001). Design of the sum-of-conforming-run-length control charts. *European Journal of Operational Research*, **132**, 187-196.

Xie, M. and Goh, T.N. (1992). Some procedures for decision making in controlling high yield processes. *Quality and Reliability Engineering International*, **8**, 355-360.

Xie, M. and Goh, T.N. (1993). Improvement detection by control charts for high yield processes. *International Journal of Quality and Reliability Management*, **10**, 24-31.

Xie, M. and Goh, T.N. (1997). The use of probability limits for process control based on geometric distribution. *International Journal of Quality and Reliability Management*, **14**, 64-73.

Xie, M.; Goh, T.N. and Xie, W. (1997). A study of economic design of control charts for cumulative count of conforming items. *Communications in Statistics - B: Computation and Simulation*, **26**, 1009-1027.

Xie, M.; Goh, T.N. and Lu, X.S. (1998). A comparative study of CCC and CUSUM charts. *Quality and Reliability Engineering International*, **14**, 339-345.

Xie, M.; Lu, X.S.; Goh, T.N. and Chan, L.Y. (1999). A quality monitoring and decision-making scheme for automated production processes. *International Journal of Quality and Reliability Management*, **16**, 148-157.

Xie, M.; Goh, T.N. and Kuralmani, V. (2000a). On optimal setting of control limits for geometric chart. *International Journal of Reliability, Quality and Safety Engineering*, **7**, 17-25.

Xie, M.; Goh, T.N. and Tang, X.Y. (2000b). Data transformation for geometrically distributed quality characteristics. *Quality and Reliability Engineering International*, **16**, 9-15.

Xie, M.; Del Castillo, E.; Goh, T.N. and Cai, D.Q. (2001a). On the monitoring of trended and regularly adjusted processes. *International Journal of Production Research*, **39**, 3641-3650.

Xie, M.; Goh, T.N. and Tang, X.Y. (2001b). A study of economic design of cumulative count of conforming control chart. *International Journal of Production Economics*, **72**, 89-97.

Xie, M.; He, B. and Goh, T.N. (2001c). Zero-inflated Poisson model in statistical process control. *Computational Statistics and Data Analysis*, **38**, 191-201.

Xie, M.; Tsui, K.L.; Goh, T.N. and Cai, D.Q. (2002). Process capability indices for regularly adjusted process. *International Journal of Production Research*, to appear.

Xie, W.; Xie, M. and Goh, T.N. (1995). Control charts for processes subject to random shocks. *Quality and Reliability Engineering International*, **11**, 355-360.

Yang, Z.L. and Xie, M. (2000). Process monitoring of exponentially distributed characteristics through an optimal normalizing transformation. *Journal of Applied Statistics*, **27**, 1051-1063.

Yang, Z.L.; Xie, M.; Kuralmani, V. and Tsui, K.L. (2002). On the performance of geometric chart with estimated control limits. *Journal of Quality Technology*, to appear.

Zhang, G.Q. and Berardi, V. (1997). Economic statistical design of $\overline{X}$ control charts for systems with Weibull in-control time. *Computers & Industrial Engineering*, **32**, 575-586.

Zhang, N.F. (1998). A statistical control chart for stationary process data. *Technometrics*, **40**, 24-38.

Index

L

M

N

O

P

Q

R

S

T

U

W

X

Z